Science for Sale

Science for Sale

The Perils, Rewards, and Delusions of Campus Capitalism

Daniel S. Greenberg

The University of Chicago Press :: Chicago and London

Daniel S. Greenberg is a journalist who has written extensively on science and health politics. He is the author of *Science, Money, and Politics: Political Triumph and Ethical Erosion* and *The Politics of Pure Science*.

The University of Chicago Press, Chicago 60637
The University of Chicago Press, Ltd., London
© 2007 by Daniel S. Greenberg
All rights reserved. Published 2007
Printed in the United States of America

16 15 14 13 12 11 10 09 08 07   1 2 3 4 5

ISBN-13: 978-0-226-30625-4   (cloth)
ISBN-10: 0-226-30625-9   (cloth)

Library of Congress Cataloging-in-Publication Data

Greenberg, Daniel S., 1931–
  Science for sale: the perils, rewards, and delusions of campus capitalism / Daniel S. Greenberg.
       p.   cm.
  Includes bibliographical references and index.
  ISBN-13: 978-0-226-30625-4 (alk. paper)
  ISBN-10: 0-226-30625-9 (alk. paper)
    1. Research—United States—Finance. 2. Universities and colleges—United States. 3. Research institutes—Economic aspects—United States. 4. Science—Economic aspects—United States. 5. Federal aid to research—United States. I. Title.
  Q180.55.F5G74   2007
  500.71′1—dc22

                              2006102639

♾ The paper used in this publication meets the minimum requirements of the American National Standard for Information Sciences—Permanence of Paper for Printed Library Materials, ANSI Z39.48-1992.

Contents

Part Three: Fixing the System

A Background Note and Acknowledgments

This book draws on a career in science journalism that began in 1961, when I joined the staff of *Science*. Since then I have been continuously steeped and educated in the ways of science through innumerable interviews, hearings, briefings, conferences and meetings, and visits to research centers.

During a decade at *Science*, I served as reporter, news editor, and London-based European correspondent. Along the way, I held an appointment as a research fellow in the Department of History of Science at Johns Hopkins University while writing my first book, *The Politics of Pure Science*, first published in 1967 (new edition in 1999 by the University of Chicago Press).

In 1971 I founded *Science & Government Report*, an internationally circulated newsletter, which I edited and published for over twenty-five years. I've also written for other publications, including the *New England Journal of Medicine, Nature, New Scientist*, the *Economist*, and the *Lancet*. And for many years, I wrote an op-ed column on science and health politics that appeared in the *Washington Post* and other newspapers. My second book—*Science, Money, and Politics: Political Triumph and Ethical Erosion* (University of Chicago Press, 2001)—was written while holding an appointment at Johns Hopkins as a visiting scholar in the Department of History of Science, Medicine, and Technology.

In focusing on the present book, I have benefited from a great deal of assistance. The Brookings Institution, in Washington, D.C., provided an appointment as guest scholar. The Robert Wood Johnson Foundation provided financial support through an Investigator Award in Health Policy Research. At annual meetings with fellow RWJ investigators, almost all academics, I have received invaluable insights on the complexities of science for sale. From among them, I am particularly grateful to David Blumenthal, professor of medicine and health care policy at Harvard Medical School, who has researched and published extensively on conflict of interest and other issues of scientific integrity. Goldie Blumenstyk, who has written about technology transfer for the *Chronicle of Higher Education* for many years, helpfully reviewed parts of the manuscript. I also learned a great deal from three of the best-informed critics of academic commercialism: Michael Jacobson, head of the Center for Science in the Public Interest; Merrill Goozner, director of the center's Integrity in Science project; and Sidney Wolfe, director of the Public Citizen Health Research Group. For the statistics that illuminate the commercialization of science, I have relied on data compiled by the National Science Foundation, the Association of University Technology Managers, the *Chronicle of Higher Education,* individual universities, and other sources, all identified in chapter endnotes. Their compilations sometimes overlap, sometimes address different aspects of the same topic, and even vary in choice of calendar year, fiscal year, and academic year. The choice of which source to use depended on the appropriateness of its numbers for particular subjects. However, given the complexities of accounting for innumerable research grants, expenditures, patents, royalties, and many other items in our far-flung research enterprise, these sources are not always in harmony; and routinely they are not current, as data collection and processing often lead to publication lags of two years or more. Nonetheless, the statistics from the various sources more or less match and are generally reflective of present-day realities. The many scientists and administrators quoted in this book were indispensable for advancing my understanding of the issues and problems of science and money. They are identified by the position they held at the time of our conversations. Finally, most important of all for this writer was the love, unerring judgment, advice, and encouragement of my wife, Wanda.

Though all of the above and others, too, provided assistance, I alone am responsible for any errors or other shortcomings in this book.

Introduction

In all cases, money drives the engine of a university. Salaries at this place are not guaranteed by anybody. Even the tenured faculty get only a portion of their support from the state, so that the pressure to bring in funds from wherever is huge. That's how you have to live: on patient care, clinical research, basic research, what have you. And I would say that under those circumstances, money talks. So the degree to which it compromises the integrity of the institution or the individual—that's a tough call. I would like to believe that very little of it contaminates us, but I think there are plenty of people who say it's very difficult to resist being kind to the person who is paying your rent. I'm not aware of any major issues right now, either involving patient care or basic research, or in terms of abusing the postdocs and the graduate students. The watchdog activities, certainly the IRBs [institutional review boards] are quite good, and the regulations are getting tougher and tougher. I'm wondering if, in fact, it's not quite excessive. I think it inhibits research by a number of people. I'm an example. I just can't do it.

Daniel Bikle, professor of medicine, University of California, San Francisco; former chairman, UCSF Conflict of Interest Task Force[1]

: : :

Don't underestimate the power of greed in the halls of science or the wholesome presence of altruism and self-respect. And don't overlook shame and embarrassment as forces for good behavior in scientific affairs.

I found these precepts helpful navigational aids for exploring the dimly lit relationship between academic science and corporate America. Coming from knowledgeable quarters, persistent alarms raise troubling concerns about ethical erosion in a great pillar of modern civilization and human well-being, the scientific enterprise. Universities are the major producers of fundamental science and large amounts of technology, and they are the training ground for the future of science, medicine, and engineering. (For convenience, I'll generally use the word "science" to encompass all three.) University-employed specialists in these fields are producers, interpreters, and custodians of precious understanding in a world where relatively few people comprehend the workings and potential of science. By the traditions of their calling, and in popular understanding, professionals in and around the sciences are expected to possess a selfless dedication to truthfulness, the growth of knowledge, and the public interest. Scientists aren't made of better moral stuff, we were long ago instructed by a great sociologist of science, Robert Merton. Rather, he pointed out, "Involving as it does the verifiability of results, scientific research is under the exacting scrutiny of fellow-experts. Otherwise put . . . the activities of scientists are subject to rigorous policing, to a degree perhaps unparalleled in any other field of activity."[2] Merton wrote those reassuring lines in 1942, another age in the practice and pace of science. And, important to remember, the policing of science, and its habits of truthfulness, has traditionally been focused on the conduct and reporting of research. The sale of science is a relatively new phenomenon, and it follows the modern ways of business, rather than the ancient ways of science.

Have today's commercial values contaminated academic research, diverting it from socially beneficial goals to mercenary service on behalf of profit-seeking corporate interests? What are the gains and losses in the visibly tightening linkage of science and mammon, and to whose benefit and whose detriment? Can academic institutions, with their insatiable appetite for money, reap financial profits from their production of valuable knowledge without damage to the soul of science and the public?

Over the past decade, press reports and an abundant professional literature have presented numerous accounts of researchers and their universities aggressively prowling for riches in capitalist territory. The congratulatory versions tell of large sums received by universities and their researchers in return for providing society with useful science and technology and new jobs based on academic discoveries. These accounts, however, are shadowed by confirmed horrific examples of researchers and administrators in renowned universities betraying scientific integrity in return for personal or institutional financial gain.

The record of sins is by now large and dolefully familiar. At the insistence of pharmaceutical-industry sponsors, research findings unfavorable to their products have been suppressed, sometimes by obliging university scientists, sometimes with the collaboration of university administrators intimidating or overriding faculty researchers. Human volunteers in clinical trials have been needlessly endangered, and some have died in reckless experiments. In conflicts over environmental health and safety, "hired gun" scientists have collaborated with industry in efforts to mislead the public and stymie regulatory agencies. The ideal of openness and collegiality for the advancement of science has always had to contend with the pursuit of personal glory. But surveys show that corporate money on campus and opportunities for commercial gain now also figure in scientists shielding, rather than sharing, their data.[3] Seminal tactics for co-opting the prestige of academic science in behalf of corporate interests were developed by the tobacco industry following the publication in 1964 of *Smoking and Health: Report of the Advisory Committee to the Surgeon General of the Public Health Service,* which launched the federal government's war on tobacco. For decades afterward, university research financed with tobacco money provided cover for the industry, even when the results confirmed the lethal effects of tobacco. In a typical instance in 1979, when an industry-supported research project linked smoking and heart disease, the industry-financed Council for Tobacco Research publicly questioned the cause-and-effect relationship and noted, "Grantees are always encouraged to publish their findings. . . . This and so much else in the medical literature just shows that we have a great deal more to learn before we can reach any solid conclusions about smoking. It may or may not be hazardous, and that's where we are."[4] Twenty years later, academic science and medical education often tolerate similar exploitation of their prestige and public respect in return for money, today mainly from pharmaceutical firms—an industrial sector that has repeatedly undermined scientific integrity and patient well-being in pursuit of profits.

The indictment of commercialized science states that research universities, despite their nonprofit status and projection of goodness, have become another form of moneymaking enterprise, and in the process they have abandoned cherished values that distinguish truth-seeking from profit-seeking. Even the hallowed U.S. National Institutes of Health, the world's greatest and wealthiest medical-research institution, has been blighted by revelations of private, profitable dealings between several dozen of its senior administrators and pharmaceutical firms—a small number in the context of the NIH's size, but nonetheless a blight.

Concerns arise from another matter, a subtle one: the deliberate infusion of entrepreneurial goals into the academic environment, so that attentively or not, university leaders, professors, and their students inhale commercial values as they make their choices and perform their work.

Note that the sense of loss and change, perhaps even of betrayal, is predicated, in both popular and professional ranks, on a touching faith in ethical purity as a natural condition in science. Whether the faith in scientific saintliness is historically justified is a separate issue. As a practical matter, however, honest science, unsullied by commercial lures and pressures or politics, is clearly an invaluable public good, whether for assessing the safety of drugs or the menace of global climate change. The legal system sensibly rejects a judge or juror with an interest in the case at hand. Should the same apply to scientists with financial connections to industry who counsel government agencies and advise journal editors and the public? Does disclosure suffice to sanitize their conflicts of interest? It is often applied in scientific affairs. But even the vaunted value of disclosure and transparency as purifiers of conflicts warrants some skepticism. Do these obligatory confessions encourage or signify objectivity and disinterested judgments? Or do they simply induce complacency?

All speak in favor of honest, independent science, even those who seek to subvert it. In defense of science's tango with commercial enterprise, a counterresponse of the "yes, but" variety comes from corporate managers, venture capitalists, entrepreneurial scientists, and university officials responsible for promoting and also ethically monitoring academic-industrial collaboration. They assert that collaboration between academe and industry is a necessity because of their complementary capabilities: fundamental discovery by university scientists and product development, production, and marketing by industry. In these dealings, the universities contend, public service rather than money is their primary goal, though the money is acknowledged as welcome and needed. And it is ardently pursued. Purity and profits are deemed compatible, and while the salesmen of academic science acknowledge that worrisome misdeeds have occurred, they insist that individual and institutional malefactors in science are rare and that safeguards to thwart them are in place or are taking hold. For certain, there is no dearth of exposure and diagnosis of wrongdoing in science and its applications. At issue—and the focus of this book—is whether proper correctives have been formulated and whether they are being effectively applied.

Distress with the present ethical condition of science can induce an unrealistic nostalgia. Up until World War II, and for some years afterward, the scientific community was far smaller, cloistered, and, by universal account, far less materialistic. Today it is a huge enterprise of immense economic, political, and cultural consequence. From all sources, annual spending for research and development in the United States exceeded $300 billion in 2004—some 2.5 percent of gross domestic product—more than the R&D spending of the European Union nations combined.[5] Academic science, which spent about $42 billion of the U.S. total, eternally has money on its mind because it is always short of money, which inspires highly organized efforts for turning scientific knowledge and skill into personal and institutional wealth. At universities throughout the country and in government laboratories, too, doing so has become a high-priority mission, even as the public-service rationale is increasingly emphasized while the profit rationale is discreetly hushed.

There was a time, not so long ago, when academic commercialism was unfashionable, even tending toward being regarded as unclean. "Who holds the patent on this vaccine?" Jonas Salk, inventor of the polio vaccine, was asked in 1955. "Well, the people, I would say," he replied. "There is no patent. Could you patent the sun?"[6] Universities currently receive close to four thousand patents a year based on research performed by their scientists and other faculty members. And when infringements or legal challenges threaten their patent income, they go to court, at considerable cost, to protect their interests, knowing that they may not win; nonetheless, they litigate to protect their revenue.

Since ancient times, truth-seeking, wherever it leads, has been enshrined as the sacred obligation of the scientist. But contemporary science is embedded in, and financed by, a society that worships money and profits and celebrates personal wealth. Ethical purity and profits are not necessarily at odds, but they sometimes are. Secrecy is an accepted tool of the business culture. Patenting and equity shares in corporate spin-offs from campus labs have become a familiar aspect of the professorial life. Inevitably, profit-induced secrecy has seeped into academic science. We should wonder how could it be otherwise, given the monetary potential of scientific information. In its organization and financing, science is not a static enterprise, nor could it be in the modern American economy. New modes evolve in the financing, management, and direction of research. The personal values of recruits to the profession also evolve as the society around them changes. Before World War II, federal financing of academic science was feared and opposed by many scientific leaders as potentially corrupting, and it was virtually nonexistent, both by their choice and intentional government aloofness from support of science. Salaries in the sciences were low, and opportunities for getting rich were rare. Philanthropic foundations and wealthy individuals financed much of science in that era, with state governments providing a share, mostly for agricultural research. To-day scientists pursue federal money and find it both "clean" and indispensable. Fear of government support has been succeeded by other concerns. Critics look askance at the proliferation of biotechnology firms founded by entrepreneurial professors, regarding this marriage of university science and capitalism as a sellout of the public interest. The union of campus and corporation is denounced as detrimental to the progress of science because of the intrusion of secrecy, normal in business but supposedly foreign to academic science. The "scientific commons"—the body of knowledge open to all scientists—is seen as threatened by promiscuous patenting, concealment for commercial advantage, and other transgressions against scientific openness, collegiality, and cooperation. However, the social and economic texture of modern science is complex. Some of the earliest critics of commercialized science now candidly acknowledge that biotech firms have overcome the traditional narrowness of academic research projects and rigid departmental structure and have evolved valuable methods of collaboration. Many of their scientists publish in the same journals that carry the papers of university scientists. "Clear your mind of cant," said Samuel Johnson, in denouncing meaningless, pious rhetoric. Good advice in all things, including our subject.

While many documented episodes support the grim assessments and confirm the dour prophecies of the critics of science for sale, my aim is to introduce some measure of skepticism about the width, depth, and especially the durability of the alleged rot. Equally important, I find reason to doubt that academic-business dealings are as economically and socially important as their boosters claim; and to the extent that they are, that the cash-and-carry model now in effect is the best way to conduct them. These issues hold repertory status on the busy, separate conference circuits that resound, on the one hand, with charges of scientific collusion in corporate misdeeds and, on the other, with celebrations of new heights in academic-corporate business deals. Fortunately, there's also an even-tempered, research-based scholarly dialogue based on fact gathering, for which I am grateful.

I hope to direct attention to the scientific community's potential for correcting serious failings that are bound to worsen in the absence of effective remedies. Though it may seem doubtful, university leaders, administrators, and scientists sometimes are educable, and when they are not, they can be shamed or coerced into behaving properly. The starting point must be recognition that much is amiss in the house of science, and, in the manner of deviant behavior, there's surely more gone wrong that we never learn about.

During many years of reporting the politics of science, I came to sense that repetitive reports about complex matters of public affairs warrant another look. In particular, I observed that many of the expressed concerns about the ethical decline of science recycled the same atrocity stories, all verified, shocking, and inexcusable, but relatively few in number, increasingly distant from the time of occurrence, and of uncertain representativeness in the big world of American science. I realized this after touching on the issues of science for sale in an earlier book that was predominately concerned with relations between science and Washington, *Science, Money, and Politics: Political Triumph and Ethical Erosion* (University of Chicago Press, 2001). Though several often-cited surveys have found a high percentage of university researchers, particularly in the life sciences, reporting linkages to industry, upon examination the nature of those links range from intense to slight to near nonexistent. Sometimes they involve a great deal of work on industrial products and sometimes no more than an annual lecture to industrial scientists or an arrangement to consult by telephone occasionally. At many universities I encountered strong concerns about ethical lapses, large and small, inadvertent or deliberate, that might besmirch the institution or a part of it. The mesh of rules and checkpoints inspired by such fears may be desirable, but they are without question a bureaucratic drag on the conduct of research. On the other hand, easily blocked misdeeds continue to flourish in relations between science and industry. For example, the pharmaceutical industry does not lack for professors willing to put their names on ghostwritten scientific papers that promote the sale of particular drugs. Scientific and medical journals have been slow to press their authors for assurances of independence and all relevant data in their submissions for publication. The embrace of entrepreneurship as a major campus-wide activity at some universities invites wonder about their values and sense of purpose.

And so, in 2004 and 2005, with the financial assistance, but hands-off policy, of the Robert Wood Johnson Foundation, I conducted an on-the-ground reconnaissance of the relationship between academic science and profit-seeking enterprise. My research technique consisted of talking—in most instances with my tape recorder visibly running nearby—with researchers, administrators, and technology-transfer specialists from over twenty large and middle-size universities and research centers. I also talked with federal regulators, officials of professional societies, lobbyists, patent lawyers, journal editors, and business executives, about two hundred in all, usually in their offices, sometimes at professional or social gatherings, occasionally in telephone interviews. For comparisons, I also talked to some of their counterparts in Great Britain. There, as in other countries, American vigor in academic-industrial relations and in the creation of university-spawned businesses is admiringly studied in pursuit of creating jobs and making money. But also in Britain, the growing intimacy between universities and industry has produced abuses, serious misgivings, and calls for corrective measures. The words of my interviewees are extensively presented throughout this book, with our conversations edited by me for brevity and clarity. The people who are quoted on substantive matters are identified by name and position, with no anonymity. To take the reader deeply into some of the complexities and subtleties of academic-industrial relations, in part 2 of this book, I have excerpted large chunks from some of the most revealing and instructive conversations. To inform myself, I have read or skimmed a fair portion of the voluminous academic, journalistic, and polemical literature on this subject. To supplement my own findings, I have quoted, with attribution, from the writings of others.

Proceeding in this fashion guaranteed failure to achieve a definitive view of what's going on out there. The subject is too big and too varied from university to university, and even within universities, to capture the whole story, which is rich in nuances, misleading appearances, hyperpolemics, self-delusions, deliberate evasions, and overlooked realities sitting in plain sight. Moreover, while misdeeds stubbornly persist, beneficial changes in the direction of upright behavior have been occurring in the last few years, in direct response to publicized abuses and some government interventions. This is particularly true of the standards and procedures for protecting human volunteers in medical experiments, an area of research with a dreadful record. My interviews left me impressed with the abundance of honesty and integrity in the ranks of scientists, qualities often obscured by the outrageous behavior of some of their colleagues and managers. In comparison to team research or broadly distributed questionnaires, the inquiring solo reporter—accumulating knowledge, clues, and intuitive feelings along the way—can assemble a revealing picture that may otherwise be unattainable. Readers must make their own judgments.

In a field that has produced jeremiads from the critics of science for sale, and many fairy tales from the enthusiasts, what follows is an explorer's report.

**Part One: The Setting
and the System**

1

Money for Science: Never Enough

We run our research enterprise primarily in a self-funding way, but not in a profit mode. So we have to work very, very hard on a competitive basis to bring in the monies to pay the people, to fund the indirect costs, which includes replacing buildings. Actually, we end up subsidizing research through other revenue sources in the medical center to keep it going. Some faculty members will be recruited here to become wonderful, superb clinicians and, in doing so, clinician-teachers. Others will be more in what we call the physician-scientist track, have very modest clinical activities, teach graduate students and basic-science medical students. And, yes, we expect them to fund almost all of their laboratory activities through external funds. Almost all faculty are expected to bring in most of their own support. We try to be very, very supportive, because anyone can have a short downtime. So we provide good, reasonable inter-grant support, but not forever.

Robert P. Kelch, executive vice president for medical affairs and CEO, University of Michigan Health Systems[1]

: : :

Most serious science in the United States is conducted in the big institutions known as research universities. The counts vary, but about fifty of them are in the scientific big leagues, conducting research and producing PhD's across at least several important disciplines; perhaps another fifty are striving to join them. On money matters, all these universities are puzzling and contradictory organizations. Virtually all describe themselves as hard-pressed financially, even as they ingest colossal sums from a variety of sources, accumulate huge endowments, and operate on enormous budgets. In 2006 Harvard's endowment reached $29.2 billion, a one-year increase of $3.3 billion, and its operating budget was nearly $3 billion. For the University of Michigan, the endowment in 2005 stood at $4.9 billion, and revenues for its three campuses—at Ann Arbor, Dearborn, and Flint—were $4.2 billion; for Stanford, $12.2 billion in endowment and a budget of $2.9 billion, plus a capital budget of $373 million; and for Johns Hopkins, $2.1 billion in endowment and a $2.4 billion budget.[2] Their begging and searching for money never stops, while most practice miserliness in drawing upon their mounting endowments. Exempted from federal regulations that require tax-exempt foundations to spend at least 5 percent of their endowments annually, most universities spend less—in 2004, 4.5 percent each for Harvard and Yale, 4.1 percent each for Princeton and the University of California. A few spend more, but very little more.[3] Despite the appearance of wealth, annual increases in tuition above the rate of inflation are a common feature of academic finance, though price-cutting deals are routinely made to bring in academically high-ranking students and other desirables, particularly star athletes and, lately, impoverished minority members who show academic promise.

With populations of students, faculty, and staff running into the scores of thousands, big universities are modern versions of the city-state. The University of Wisconsin–Madison, for example, lists 41,000 students, 2,250 faculty members, and 7,000 professional and administrative employees.[4] Arizona State University, with 61,000 students in 2005, aims to enroll 95,000 by 2020 and double its research budget—which stood at $183 million in 2005—within three or four years.[5] Employing persistent and sophisticated dunning methods, fund-raising campaigns run continuously in academe, with the billion-dollar mark, or more, often set as an inspirational goal at the mega-institutions. In 2003–4, gift collections totaled $540 million at Harvard, $524 million at Stanford, and $385 million at Cornell.[6] The *Chronicle of Higher Education* periodically reports the progress of the twenty to twenty-five universities running billion-dollar, or more, fund-raising drives.

Universities possess their own security staffs, residential housing, schools for children, health facilities, newspapers and TV stations, theaters, places of worship, recreation facilities, and even courtlike bodies for judging infractions by both students and faculty. Like sovereign governments, they hold elections and they levy taxes, known in their context as tuition. Universities are increasingly innovative in developing relationships outside their boundaries, with local and national business firms, surrounding communities, and the federal and state governments. And they're constantly tinkering with their programs, with student internships, study abroad, combined undergraduate and graduate studies, and course offerings without bounds. A hoary legend of academe has it that Woodrow Wilson, in his frustrating pre–White House years as president of stodgy Princeton University, complained: "It's easier to move a cemetery than it is to change the curriculum." True at a few schools today, but not many.

In many settings, such as Johns Hopkins University, in Baltimore, the vast Boston-Cambridge higher-education concentration, the University of Michigan–Ann Arbor, and several University of California campuses, universities are the biggest or among the biggest employers and spenders in the region. They look rich, even while maneuvering around the operating deficits that chronically plague many of them. Small colleges occasionally collapse and disappear for lack of money and students. Following publication in 1910 of the *Flexner Report* on medical education, scores of substandard, for-profit doctor-training mills went out of business. In a rare modern occurrence, two small-ish, free-standing medical schools in Philadelphia merged in 1993 to become MCP Hahnemann University, which was merged into Drexel University in 2002 as the Drexel University College of Medicine. In 2004 Britain's University of Manchester Institute of Science and Technology disappeared through a merger with the University of Manchester, to form a supersize university with global ambitions of scientific and academic greatness. Except for such events, which are extremely rare, big universities possess a unique quality: they are immortal. Founded in 1088, the University of Bologna is still there. So are Oxford and Cambridge (twelfth and thirteenth centuries), Harvard (1636), Yale (1701), Princeton (1746), and hundreds of others from prior centuries. Empires dissolve. Nations split. Religions fragment. Dissolved or absorbed in mergers are numerous corporate behemoths of yesteryear: Studebaker, Packard, American Motors, Pan Am, TWA, RCA, Bethlehem Steel, Wang, Digital Equipment—all prey to capitalism's unsentimental "creative destruction," or mismanagement. In contrast, universities survive, and expand, situated in their own specially insulated, nurturing economy, based on government support, philanthropy and private gifts, unrestrained pricing for coveted enrollment slots, and freedom from taxation. Like potentates of yore, university leaders regard expansion as an imperative. The surrounding world encourages their ambitions. As societies grow in complexity, needs expand for the training, education, and professional certification provided by universities. Hundreds of institutions have metamorphosed from small seminaries and teachers and agricultural colleges to four-year arts and sciences colleges and then onward to the sprawling, academically diversified, PhD-granting universities that are the flagships of modern higher education. Aggressive pursuit of growth—for students, money, programs, buildings, even acreage—is an ingrained trait of modern universities. A modern expansionist tactic creates satellite campuses in their region, and even farther beyond, including in other nations. The top universities have become notorious for student-recruitment wiles that emphasize, and frequently exaggerate, their exclusivity; many of the lesser institutions, however, scrape to fill their classes, often employing consultants skilled at ferreting out the needed students. The infusion of corporate culture into the modern university is reflected in help-wanted ads for skills far distant from teaching, research, and the traditional tasks of academic administration. Thus, not atypically, the University of Idaho announced that it was seeking an assistant vice president for marketing and strategic communications who will be

> responsible for managing the university's image, media relations, publications, advertising, world wide web communications, university branding, corporate identity, communications planning, presidential communications, event promotion, crisis communications and market research.[7]

In Greatest Need of Money

The costliest parts of the modern university are the science, engineering, and medical components. Like the very rich, they need more money because they spend more money to satisfy their expensive tastes. They excel all other academic fields in drawing money to the university, wads of it from government agencies specially created to finance them. For bringing glory and public attention to a university, the science-related departments are exceeded only by the athletics department. But, unlike athletics, which are often a money-losing proposition, the sciences also bring universities a kind of income little known to the outside world but extremely appealing in the stringent environment of university finance: reimbursement for indirect costs, also referred to as overhead costs, which are distinct from the readily visible direct costs of research, such as equipment, supplies, and salaries.

Indirect costs are the nonscience expenses that a university incurs from the presence of research on its campus, such as security guards for the laboratories where the research is conducted, depreciation, library services for the scientists in the laboratories, and financial staff to track their money, write their paychecks, process orders for their equipment and supplies, and pay the electric bill. To maintain the fiction of a government-university financial partnership in research, Washington— the main source of money for academic research—pays only part of the claimed indirect costs, leaving the balance to the university as its share of the tab. Nevertheless, for each of the big research universities, with receipts of hundreds of millions of dollars a year in federal research funds, reimbursement for indirect costs provides additional scores of millions of dollars. The computation of these costs has developed into an arcane accounting specialty, but the outcome is a large helping of federal money atop the money specifically destined for the laboratory. The amounts awarded for indirect costs ostensibly range from about 30 to 90 percent or more of direct costs, depending on the type of research and the characteristics of the institution. But with a variety of exceptions and caps written into the rules, the take invariably runs below the stated rate, averaging out to about 30 percent of direct costs. Even so, more research means more money coming in for indirect costs, a link that adds allure to acquiring federal grants. For the federal agencies that are the financial mainstays of academic research, the slice of their budget taken by indirect costs annually totals billions of dollars. In 2005, when the NIH budget totaled $28 billion, $5.9 billion of that sum was expended on indirect costs, that is, costs incurred outside the laboratories but on the campuses of the NIH's academic beneficiaries. In that year the budget of the National Science Foundation stood at $5.4 billion, of which $980 million went to indirect costs.[8] Underlying this academic-government partnership in science are strong political and cultural convictions that science is beneficial for the country, and that universities are a good place to do science because they train new generations of scientists and are, or should be, bastions of independent expertise for dealing with public problems.

While the reality of those expenses generated by science on campus is not disputed by the government, universities contend that the indirect costs are burdensome and the federal funds to cover them are insufficient. But in the hardscrabble world of university finance, they accept what they can get. Private foundations recognize that indirect costs also arise from their philanthropy, but they generally hold the level down to 15 to 20 percent. Even at these low rates, they do not lack academic takers.

The humanities can get along with books, a modest amount of travel and conference expenses, graduate stipends, and relatively simple computer facilities. Washington knows that and gives the National Endowment for the Humanities a mere $150 million or so a year, while the university-based medical and physical sciences annually receive over $30 *billion*. From this sum, a few crumbs are provided for the social sciences, but the great bulk of the money is for "real" science: biology-related research, physics, chemistry, astronomy, engineering, plus their interdisciplinary offshoots and a few others. Humanists helplessly rail against the disparity, as in a recent essay in the *Chronicle of Higher Education*:

> Never before has there been such inequality among the disciplines and schools that make up a university. . . . Disciplines like history, sociology, philosophy, the visual arts, and literature were once seen as the heart of the university, respected as sources of wisdom about the human condition. But over the last 10 years, faculty members in those disciplines have become the poor relations of the hard-science powerhouses, who have higher salaries, greater abilities to hire, and better chances of attracting better students, who will themselves leave with better jobs.[9]

The so-called real sciences basically operate in two realms, both heavily financed by their federal patrons. First there are "user" facilities for the huge and costly equipment of modern science, such as particle accelerators, synchrotrons, radio and optical telescopes, along with the panoply of hardware and services needed for oceanography, space research, and other fields of "big science." Scientists from various universities compete to use these facilities, which are usually managed by university consortia. Then there are home-campus facilities, roomy halls of costly equipment constantly threatened with obsolescence by new and, invariably, costlier equipment. A couple of million dollars to equip and staff a laboratory for a new hire is not unusual. While chatting with a scientist in a characteristically crammed office/laboratory, he cautioned me not to bump against a piece of equipment close by the only open space for a visitor's chair. "That costs $150,000," he said. To stretch their budgets, government research agencies request scientists to share expensive equipment, especially apparatus that is used intermittently. But scientists like to have their own equipment, nearby and available when they want it. If the money is to be had—and often it is—they'll acquire their own, even if the very same equipment down the hall is mostly idle.

The billions in federal funds for academic science do not suffice. They amount to only 60 percent of the money that universities spend on science, with the balance mainly coming from tuition, gifts, income from endowment, licensing fees for patents, and wherever other money can be scraped up.[10] Industry and state governments also help finance academic science, but as we will closely examine later, these two sources pay a surprisingly small share of the costs. Despite, or because of, the mammoth amounts required for running a major university, money is a repeatedly declared universal dearth, as any reading of the journalism of higher education will confirm, with its constant reports of multiyear fund-raising campaigns in the billions of dollars, tuition increases, and recurring budget crises. An after-dinner speech quip jokes that "a university president is someone who lives in a big house and begs for money." Academic cup-rattling endeavors are typically sophisticated in style and segmented according to the potential for the yield, as indicated in a recent help-wanted advertisement by Princeton University. Announcing two openings in its Office of Leadership Gifts, Princeton stated, "The successful candidates will be responsible for gifts at the six-to-seven figure level by identifying, cultivating, soliciting, and managing a portfolio of approximately 200 alumni, parents and friends of Princeton University in the Northeast."[11]

Pay Scales: The Low and the Lofty

The popular image of the jet-setting scientist is realistic. Funds for holding conferences and grants to attend them are often included in government research awards. There are many grants and there are many conferences, often in interesting, faraway places, even if the proceedings are dreary, as veteran attendees often complain. The average laboratory scientist, however, is an unlikely exemplar of personal wealth, capitalistic instincts, or sumptuous living, beyond travel and a few other modest perks that come with the job. Moneymaking has never ranked high as a motivation for a scientific career. In today's economy, academic science is one of the least remunerative and most uncertain career choices on the professional landscape, despite the flocks of millionaires and occasional billionaires spawned by science-based companies.

Consider the options for a student graduating from college: three years of law school or a year or two in business school will provide a reasonable assurance of a comfortable and upward income. The cost, perhaps, is mind-numbing work, but that's the deal, and recruits are not lacking. In contrast, the prescribed apprenticeship for a career in modern science is far longer, with a substantial percentage of poor outcomes for many aspirants. In the science sector, homegrown recruits are lacking to the point that foreign students fill half or more of the graduate slots in science and engineering; fortunately so for American science, since graduate students perform the day-to-day work of modern research. Free tuition and stipends—$15,000 to $20,000 a year—help ease the way through graduate school for science students. But the route through undergraduate, graduate, and postgraduate scientific training is a preparatory marathon that brings many scientists to age forty or so before—with rare good fortune in fierce grants competition—they can achieve the prized status of principal investigator, which usually means a lab of their own, rather than working for another scientist. The proportion of NIH grants awarded to scientists under thirty-five has been going down, from 23 percent in 1980 to only 4 percent in 2001.[12] Many of the tenured statesmen of science puzzle over the declining interest in scientific careers among young Americans, oblivious that their beloved profession is a poor economic choice for the young in today's job market. Some see the problem clearly. Bruce Alberts, a molecular biologist who spent much of his career at the University of California, San Francisco, warned that "we all think it's a disaster for the future of science, because nobody can count on the young people if they're going to have to wait till they're forty for their own ideas to be tested. And even more broadly, it's a great damper on innovation, because by the time you're forty, you're not going to be very innovative, on average." Alberts, who completed twelve years as president of the National Academy of Sciences in 2005, noted that "my generation had independence at twenty-seven, twenty-eight, not forty." With dismay, he told me:

> Basically, my conclusion after looking at all this stuff, both at UCSF and at the academy, the free market doesn't work. That left unguided, you end up with a system that doesn't work well. All the old people get all the money from the federal government, for example, and all the labs will be huge, where the old people get the money and the young people have to go and work in an environment, a lab of fifty people, in order to get support. And this is not the right environment. This is not the way to get good science. And yet that seems to be the natural way that things evolve, given the system we have, which relies on competitive grants. The best people at it are all the

people who got all the experience. It's talent getting money. And a review system that always looks at preliminary results, which means that if you're not doing a safe thing starting out, you have a disadvantage. And a university system which could look at "how could I maximize my indirect costs? With one professor and a lot of research." The whole incentive system seems to be set up to give the results that we get. And it's not surprising that the growth of this increasingly aged, independent investigator group has been occurring. What's surprising to me is that nobody was noticing for ten years. Suddenly we wake up to it. It was a problem, and extrapolated for another ten or fifteen years, you'll be fifty years old before—This is ridiculous.[13]

Scientific societies and various health-related philanthropies provide grants and other financial assistance for outstanding beginners, but here, as elsewhere in science, the federal government dwarfs these and all other sources of money, and thereby sets the reigning conditions. From the billions they distribute, why can't the managers of the federal grant system set aside funds for promising youngsters? The answer derives from the eternal scarcity of money for science in tandem with the bureaucratic rigidity of the scientific enterprise. In the 1990s the NIH made a halfhearted attempt at financial affirmative action in behalf of young investigators, reserving special grants of $70,000 apiece for them at a time when the standard grant was $160,000. The smaller grants foundered in a woeful combination of circumstances: the stigma of specially set-aside money, the inadequate level of funding for performing top-notch research, and the innate conservatism of the review panels. After six years, the new grants were abolished and the money-awarding process reverted to the free-for-all competition that is historically embedded in the economy of science.[14] In 2005 the NIH announced yet another plan to assist the young by speeding up review of their grant applications—normally a nine-month process—so that failed young applicants could revise and resubmit in time for the next deadline. NIH director Elias A. Zerhouni said that the loss of rejected young grant applicants to other professions is "the one thing that keeps me up at night." [15] The prescription that ensued offered little for insomnia. In recent years, about 5,000 PhD's have been annually awarded in the biological and biomedical sciences, plus another 3,000 or so in the health professions and related clinical sciences.[16] In 2006 the NIH announced another youth program, optimistically titled Pathways to

Independence Awards, which would spend nearly $400 million, spread over the next five years, in special grants of up to $90,000 to $240,000 each year for 150 to 200 young scientists. Why such a relatively small response to the greatly deplored neglect of young scientists? Part of the answer is the eternally short money supply, which curtailed the scope of the rescue effort. The rest is to be found in the strict insistence on peer evaluation of grant proposals, even at the cost of rejecting newcomers to the profession. At the heart of the system is mandatory screening of all grant applications by expert panels, a process that supposedly keeps quality high. By invoking the prestige of scientific approval, it's also a barrier against congressional pork-barrel intrusions into the award of research money, which scientists, understandably, want to control themselves. In the rigidly enforced competition for scarce funds, experienced scientists tend to win higher ratings than unproven young scientists, and with experience comes greater savvy about navigating the bureaucratic perils of grant land. NIH and NSF administrators are proud of their competitive systems. They privately boast of turning down jaded Nobel laureates in favor of energetic, promising young scientists—though apparently not often.

Even as they lament the consequences, the managers of research favor the present against the future, to the detriment of the young. Budget pressures are the customary alibi: Give us more money, and the young will be taken care of too, they assert. But that threadbare assurance draws skepticism from influential scientists outside the NIH orbit. John H. Marburger III, the director of the White House Office of Science and Technology who serves as the president's science adviser, noted that even with the doubling of the NIH budget, "very little of the money is going to younger investigators. . . . I think the pattern of grants needs to be [reviewed]. The idea that despite the fact that there's been a massive increase in the funds available, the age at which a new investigator in biomedical research gets their first RO1 grant now exceeds 40 years. Does that make sense?" [17] It doesn't. But the customs of science, especially in the distribution of money, are difficult to alter.

The Scientific Proletariat

Though it always yearns for more, the university system is flush with money for many purposes, but except for a few superstars, salaries for teachers and researchers are low on America's professional pay scales. Salaries for medical faculty are better than most because of revenues derived from clinical practice plans at university-affiliated hospitals.

Heads of clinical departments—and, in particular, chairs of surgery—
are recompensed at football-coach levels, reaping million-dollar-plus
packages from the flood of health-insurance money coursing through
their institutions. However, even these academic-medical plutocrats,
who are relatively few in number, are minimum-wage workers in comparison to the royalty of big-league sports, entertainment, industry,
law, and business. Academic pay of $200,000 to $300,000 a year—
often the lure in a hiring raid on another campus—is so out of scale as
to make headlines. In one of the most widely reported raids of recent
times, for a salary "reported to be more than $300,000 a year, not
counting benefits," Jeffrey D. Sachs, a celebrity economist, was hired
away from Harvard to Columbia. The *New York Times* headlined the
catch with "Columbia Gets Star Professor from Harvard." [18]

Existing in its own economic zone, modern academe is a financial
mosaic of impoverishment and middling status, plus a bit of relative
affluence in the presidential suite and the aforementioned medical
locales. The pay of janitors and other campus menials is commonly
so low that right-minded students occasionally stage demonstrations
to win them a "living wage." Academic administrators usually cave in
quickly to the embarrassing spectacle of students taking up the cause of
the university's poverty-wage workers. Underpaid teaching assistants
and postdoctoral fellows flirt with unionization in their quests for a
living wage. But, with a few exceptions, they are beaten back by the
administration's pious insistence that unionization is inappropriate for
academics. Adjuncts are increasingly the stoop labor of academe, good
enough to teach undergraduates, for a few thousand dollars per course,
but not good enough for the status, pay, and security that come with
full-scale appointments.

The huge sums flushing through academe are not evident in the
smallish, cluttered professorial offices that are de rigueur on rich and
poor campuses alike. That tiresome line about academic politics—being so vicious because so little is at stake—is more than a wisecrack.
Pay scales for the rank and file of both starting and senior professors
are modest in comparison to the pay of successful, even mediocre, professionals and businesspeople in the profit-seeking world. Sartorial chic
is not abundant at faculty meetings, nor are automotive head turners
plentiful on campus, except in the student parking lots. For a full-time
professor at a private doctoral university, the average salary in the academic year 2004–5 was $127,214; at public institutions, which usually
lag behind private schools, it was $97,948. The average salary for all
professors, in all disciplines, at large and small universities, private and

public, was $68,505.[19] On academic pay scales, science professors generally fare better than the aggrieved humanists, among whom English and literature teachers averaged $54,747. But the scientists rank lower than many other professors, particularly those in the professional disciplines. Average annual salaries in 2003–4 for law faculty, $109,478; for engineering, $84,784; for business and marketing, $79,931; for the physical sciences, $67,186; for the biological and biomedical sciences, $63,988; for psychology, $62,094; for mathematics and statistics, $61,761.[20]

The academic caste system provides laboratory space for nonfaculty researchers, but only if they bring in their own financial support—known in the trade as "soft money," usually from government research agencies; in some instances, because of special talents, they are hired to work under tenured researchers. These appointments generally run from one to five years and expire when the money runs out or a particular project is completed. Nonfaculty posts rank below both tenured professorships and appointments that may lead to tenure—so-called tenure-track positions. In a poignant effort to elevate their prestige, nonfaculty researchers at Columbia University have petitioned for titular upgrades that their counterparts hold at other research universities. They wanted to shed their current range of titles—senior research scientist, research scientist, and associate research scientist—and replace it with higher-class nomenclature: research professor, research associate professor, and research assistant professor. Having professor in their title, they've argued, would improve their competitive position for grants.[21] Through a process of self-selection, academic science is not heavily populated with wealth-seeking individuals.

The professorial life provides many compensations and satisfactions, such as secure employment for the tenured, intellectual challenge and collegiality, a sense of social usefulness, opportunities for travel and moonlighting, long vacations, and paid sabbaticals. While administrators, and faculty members who aspire to administrative positions, are heroically overbooked with end-to-end committee duties, meetings, and deadlines, the rank and file of academe enjoy a good deal of independence and free time. Teaching duties at most universities occupy two thirteen- or fourteen-week semesters per year, a schedule that provides ample time for the internecine bickering and crankiness native to all campuses. Scientists actively engaged in research are usually more hard-pressed, since long hours and weekend work, outside of their classroom obligations, are often required for setting up and tending experiments. This is especially true of the young, who have a few short years to make their mark or drop out of the fierce competition for grants. Nonetheless, the scientific life still provides more personal freedom and satisfaction than the ordinary nine-to-five office routine. Academe is an attractive workplace, with many more well-qualified aspirants than jobs. Department chairs on many campuses and in many disciplines told me that the announcement of a faculty opening brings torrents of extremely well-qualified applicants—the products of the rapid, nationwide expansion of PhD programs in recent decades. But big money is not the lure. Many freshly minted graduates in law, business, and other fields outearn their former professors.

We may infer from this financial-cultural sketch that academic researchers are not driven by the profit motive. Or that because of the stringencies of the academic life, they are especially alert to opportunities to supplement their incomes from deals with industry and other forms of moonlighting. The spectrum of attitudes includes both, ranging from professors who are famously entrepreneurial to the researcher who donates an annual corporate speech honorarium to charity, simply to avoid any implication of financial motivation or corporate taint. But contrary to many breathless reports, few academic scientists are on the lookout for business opportunities, even on campuses rich in federal research funds and abundant research findings that might be spun into gold. Marvin G. Parnes, associate vice president for research at the University of Michigan, gave me his perspective:

> It's a small minority of faculty who are really eager and positioned to work aggressively with industry and do tech transfer. And I'd say the majority of faculty might give it some passing thought. If they think something they've developed or discovered were truly interesting and helpful, they might want to disclose it [for patenting]. And we've tried to educate faculty to disclose inventions and consider working in technology transfer. And a lot of them are willing to do some. But for many faculty, they are absorbed in their research, teaching, mentoring, and many of them don't want to get involved in what they view as another kind of work.[22]

On the other hand, the potential for scientific riches and the availability of assistance for patent seekers inflames commercial expectations—realistic and fanciful—among some academic scientists. Joel Kirschbaum, director of the Office of Technology Management at the University of California, San Francisco, told me of "inventors coming to us expecting us to greet them with open arms and file that patent application. But the criteria that we use to assess the business opportunity tell us that about 25 percent of these inventions are worth pursuing." Patenting is expensive, with costs ranging from $15,000 to $50,000 apiece, depending on the complexity of the invention and the need to dig up the prior history of similar inventions to satisfy the criterion of novelty in a patent application. "We're turning away 75 percent of these people," Kirschbaum said.[23]

The Boom at the Top

There is a standard exception to academe's financial aridity, and because it is situated at the top, it broadcasts a paean to money as a measure of success and importance. Over the past decade or so, the financial rewards and perks in the executive suites of universities have galloped to lofty heights, so much so that long after stepping down as president of Harvard, Derek Bok puritanically denounced "chauffeured limousines and out-of-scale salaries" for university heads. Bok, who temporarily returned to Harvard as interim president after Lawrence Summers resigned in 2006, chauffeured himself in a VW during his presidency, 1971–91; Summers rode in a limo, with license plate number 1636, the year of Harvard's founding. Warning, in 2002, that academic materialism was running loose, Bok declared that "lavish salaries for campus CEO's will only tend to make the problem worse."[24] Yet the median pay of presidents and chancellors at doctoral-granting universities, while far above professorial levels, are trifling by Fortune 500 standards, a mere $280,880 in 2004–5.[25] But note, that's the median. On a rising number of campuses, the presidency is gilded with greater amounts of money plus perks that resemble the imperial trappings of the corporate world. For both upward-striving and eminent schools, superior remuneration for the peak office is increasingly considered a necessity and is a source of pride. At several universities, total presidential compensation is now near or above the million-dollar mark.

Fifty heads of private universities received $500,000 or more in salary and benefits in fiscal 2004, and five surpassed $1 million, according to the *Chronicle of Higher Education*. In academe's private sector, a sampling of the high fliers in 2003–4 shows total compensation of $1,326,786 for E. Gordon Gee, Vanderbilt University; $1,253,352 for John R. Silber, Boston University; $939,346 for Shirley Ann Jackson, Rensselaer Polytechnic Institute; $897,139 for John E. Sexton, New York University; $895,774 for William R. Brody, Johns Hopkins University; $858,499 for Harold J. Raveché, Stevens Institute of Technology; $814,177 for Benjamin Ladner, American University; and $802,731 for Kenneth A. Shaw, Syracuse University.

State universities tended to lag behind private institutions in rewarding their chief executives but still attained high levels relative to professorial pay. In 2005–6 twenty-three state university presidents received total compensation of $500,000 or more; six were in the range of $600,000 to $699,000, and two were between $700,000 and $799,000. Topping the state list was $724,604 for Mary Sue Coleman, University of Michigan system; $693,677 for Mark G. Yudof, University of Texas system; $688,406 for Carl V. Patton, Georgia State University; $625,000 for Richard L. McCormick, Rutgers University system; $602,000 for Mark A. Emmert, University of Washington; $587,106 for G. Wayne Clough, Georgia Institute of Technology; $578,394 for Michael M. Crow, Arizona State University; $571,305 for Mark A. Nordenberg, University of Pittsburgh system; $565,090 for G. Jay Gogue, University of Houston system; $562,031 for Karen A. Holbrook, Ohio State University system.[26]

These amounts did not include typically generous reimbursement for expenses. At the presidential level, use of a car, or expenses for a car, is typical, and free housing is standard. Columbia University recently refurbished its presidential residence—also used for official entertaining—at a reported cost of $23 million.[27]

In the otherwise modest, often austere academic environment, additional presidential perks abound, including lucrative memberships on corporate boards, a low visibility but important link between academe and industry. Universities have traditionally invited corporate chieftains to their boards, partly for their managerial acumen, but also to impress upon them their need for money and to soften them up for corporate gifts. Corporations have reciprocated with board appointments for university presidents, whose aura of nonprofit rectitude is good for public-relations appearances, if not for much else in the corporate world. Corporate board memberships customarily provide $50,000 to $100,000 apiece for four meetings per year. In 2004 Rensselaer's Jackson served on the boards of directors of six corporations: AT&T, FedEx, Marathon Oil, Medtronic, Inc. (a medical technology firm), Public Service Enterprise Group (an energy firm), and U.S. Steel Corp. In the same year, Vanderbilt's Gee served on the boards of Dollar General Corporation (a retailer), Hasbro, Inc. (a game and toy manufacturer), and Limited Brands (a clothing retailer whose holdings include Victoria's Secret). Pennsylvania's Judith Rodin served on the corporate boards of the AMR Corporation (a holding company that included American Airlines), the Comcast Corporation, and Electronic Data Systems Corporation.[28] The *Chronicle of Higher Education* estimated Rodin's compensation from corporate sources at "as much as $403,900," plus "$56,860 in travel on American Airlines."

"The Million-Dollar President Soon to Be Commonplace?" the *Chronicle of Higher Education* asked rhetorically in its 2006 survey of academe's presidential salaries. Noting that public universities were catching up with their private counterparts, the *Chronicle* reported that "the number of public-university presidents earning $500,000 or more has nearly doubled in the last year, from 23 to 42." Heading the list was the president of the University of Delaware, David P. Roselle, with salary and benefits totaling $979,571, plus the customary use of a house and car, and "at least" $32,500 from a board membership, plus stock options and shares. "Still," the article noted, "pay rates at some public institutions are too low for them to hang on to their skilled leaders. The three major state universities in Iowa have lost eight presidents in the last 19 years to other universities that generally paid significantly more. Now Iowa is looking at how it can become a more serious player in what is becoming an arms race for the top university presidents." In 2002 Mary Sue Coleman moved from the presidency of the University of Iowa to the presidency of the University of Michigan system. Iowa's presidential compensation currently totals $309,250; Michigan's stands at $724,604.[29]

In the faded folklore of academe, austerity was the accepted trade-off for the special benefits of the academic life. Modest improvements have occurred for the rank and file. But at the top, big bucks and limos signal the arrival of a different culture. Bok's fears have not evoked sympathy among the salary-setting powers of academe. They observe that corporate America long ago passed the million-dollar mark for executive pay and has since gone on to multimillion-dollar compensations, even for failed or disgraced executives, and in a handful of cases, billion-dollar bonanzas. The criteria for presidential success in academe were summarized in a *New York Times* report of the unexpected resignation in 2005 of Cornell president Jeffrey S. Lehman after only two years in office: "Mr. Lehman said he was proud of his tenure as president and pointed to a 17 percent increase in applications this year, record fundraising and a significant increase in coverage by the news media."[30] The head of the Council for Advancement and Support of Education points out that even when a president "leaves in the midst of a scandal—the one thing that is pointed out is how that person was able to increase the endowment."[31] Some characteristic measures of presidential failure are revealed in the furor that arose at American University in 2005 upon disclosure that AU president Benjamin Ladner was living lavishly at the expense of the Washington, D.C., school, whose endowment, $271 million, put it low on the list of academic wealth. Ladner, however, was well compensated by big league standards, with pay and benefits totaling $881,696 in 2005.[32] But there was more. Among his self-prescribed, university-financed comforts were a twelve-course engagement luncheon for his son, prepared by Ladner's personal chef; European holidays; and maintenance of a second home, in Maryland. Ladner's reluctant departure from the presidency was assisted by a $3.7 million severance package, which led to the resignation of four indignant trustees. Decrying the "platinum parachute," they expressed their minor and major concerns: "Because of Ladner's behavior, an ethical cloud hangs over the university. The school has become the target of jokes and criticism. *And most importantly, several significant donors have withdrawn their pledges* [italics added]."[33]

The Scarcity Economy of Research

Despite affluence at the top, and the billions for research flowing from Washington to campuses throughout the country, financial uncertainty and anxiety are chronic conditions in academic science. Scientists, in most circumstances, depend heavily on periodic, but never guaranteed, grants of government money for research. Though laboratory projects generally move slowly and uncertainly, taking years, if ever, to reach fruition, the great majority of government grant durations are for only three or four years, whereupon the recipients must compete again for further support. "Writing grants"—shorthand for composing a proposal for research money—is commonly said to require several months of concentrated effort that leaves little time for research. Grant applications weave together supplication, descriptions of technical prowess, and identification of neglected problems of scientific importance—in the opinion of the applicant. To help their money-seeking scientists, many universities employ grant writers, professional wordsmiths with reputed skills for winning the hearts of the scientist-judges.

The pursuit of money is the Sisyphean chore of American modern science, laden with cautionary tales, speculations, and anxieties about the unpredictability of the awards system and the gap between its prescribed and actual workings. Among the cited risks of the process is theft of a good idea by a member of a peer-review panel, a deceit forbidden in the regulations of the policing agency for government-supported medical research, the U.S. Office of Research Integrity (ORI). Fabrication, falsification, and plagiarism are enshrined as the fundamental sins of science. Keeping up with the evolution of lab-coat crime, ORI has expanded its definition of plagiarism to include theft of ideas and information in the process of "reviewing research" in applications for grants and papers submitted for publication.[34]

A standard item of scientific folklore holds that a sure route to grant success is to propose a project that you've already quietly completed, so that ample data can be presented to enhance the plausibility of your hypothesis. The granting agencies periodically examine their peer-review systems and usually conclude that the award procedures are too cautious, too conservative, and must be prodded to gamble the government's money on long-shot projects that might pay off with scientific breakthroughs. Several years later, after "reforms" have been instituted, the next round of review is likely to reach the same conclusions. The common defense of peer review borrows from Churchill's wry salute to democracy: "the worst form of Government except all those other forms that have been tried from time to time."

The Scientific Version of Parkinson's Law

Caution in awarding research money arises from scarcity, no matter how much Washington pours into the accounts for academic science. Except for the never-never land of military appropriations, medical research is traditionally the easiest sell on Capitol Hill, where bipartisan supporters, urged on by patient-advocacy groups focused on particular diseases, champion the NIH. Alone in the $2.5 trillion–plus catalog of federal programs in recent years, medical research remained politically sacrosanct, until it got crimped by the costs of war and health care and cries of neglect from physicists and other nonmedical researchers. No legislator ever angered anyone by voting money for cancer research. Ideological-theological battles persist over anything involving reproductive biology, embryos, embryonic stem cells, and fetuses, but apart from that, support of medical research is uniquely unanimous in Congress and popular with the public.

The news media serve up volumes of health-related news, much of it in special weekly newspaper sections teeming with health advice, reports of medical breakthroughs, and personal accounts of bouts with disease and experiences with good and bad doctors. Extravagant prophecies of new therapeutics from the Human Genome Project and other gene-oriented research add political heft to the contemporary politics of biomedical research. From the early days of the overpromised "war on cancer," fears that disappointing results would erode political support for medical research have proved unfounded. When cures are not forthcoming, the response is to spend even more. Between 1998 and 2003, in an atmosphere of drastic restraints on domestic federal spending, Congress nonetheless doubled the budget of the National Institutes of Health, to an annual total of $28 billion, a sum far greater than the combined biomedical research spending of all other nations. As a result, the NIH receives half of all federal spending for nonmilitary research, while space research, physics, and other disciplines have either lost ground or experienced little growth. Even so, the NIH soon returned to Capitol Hill, pleading that funds remained inadequate to accommodate the worthy proposals from universities and medical centers across the country.

At work was the science variant of Parkinson's law: Research expands to absorb the money available for its conduct. As was the case pre-doubling of the NIH budget, after the winnowing out of glaringly low-grade proposals, funds were available for about only one-fifth of new applicants whose proposed research projects were judged deserving of support. Excellent but not good enough to receive the government's research money is the grade annually awarded thousands of research proposals. Risk and disappointment are built into the financial system of science, feeding a mood of adversity among university administrators, research managers, scientists, and graduate students. Along with writing grants, scouting for research money in Washington is another ancillary profession spawned by the research enterprise. Most big universities maintain Washington offices or regularly send their grant specialists to visit the capital city to sniff for the whereabouts of the money.

In university science departments, newly recruited young scientists are commonly given the university's own precious start-up money for research, and then, after a year or two, are on their own in the competition for grants from outside sources. In effect, they hold a license to hunt for money rather than a secure salaried position plus funds for research. Failure to succeed in the grants derby can terminate their university appointment, leaving them vocationally stranded after a decade or more of career preparation. Without grant money, they don't even rate a nonfaculty research appointment. Industry is a haven for these rejects, but in the pecking order of the sciences, even with higher pay it's considered second best. In the scientific culture, professors groom

their best students to be professors and principal investigators, ever
enlarging the ranks of those in need of research money.

The Academic "Arms Race"

Recent times have been difficult, but over the long span, government
money for science has usually increased annually, initially fed by Cold
War determination to beat the Soviets and then by political and public
faith in the economic and curative potential of research. But more is
never enough, given the open-ended nature of scientific research—each
step suggests many more—and the deep pride that universities take in
winning recognition for great scientific accomplishment. Goaded by
the annual *U.S. News & World Report* rankings of academic standing,
dubious as they are, university leaders have plunged into an academic
"arms race" of superstar acquisition, laboratory expansion, and mega-
lomanic proclamations. In this competitive restlessness, whole research
groups sometimes move from one university to another. In 2005 a
forty-member research team in child psychiatry, including eight professors, moved from the School of Medicine at the University of Chicago
to the University of Illinois at Chicago, taking with them $10 million
in grants. At about the same time, the University of Chicago School of
Medicine snared thirteen professors from Johns Hopkins, who arrived
with $30 million in grants.[35] Grant money travels with the recipients,
but also such moves are usually sweetened with additional money from
the winning university. Discretionary funds now figure in the budgets
of some public universities to ward off poaching of their academic stars
by private schools, which are generally richer and unencumbered by
state salary restrictions.[36]

   Academic grandiosity, usually involving scientific research, is publicly advertised by upward-striving universities: "Quinnipiac University
is developing and implementing a bold and far-reaching Strategic
Plan for Academic Excellence and National Prominence," including
"new faculty with superior teaching and research ability," proclaimed
an advertisement in the *New York Times*.[37] Readers of the *Chronicle of Higher Education* were informed that "as part of our five-year,
$75 million Academic Investment Plan, Northeastern University is
building research strength in four fields of great importance to the welfare of our society."[38] Grand academic ambitions were declared to the
north, where the University of Alberta "aims to be one of the world's
great universities, by transforming learning, leading ground-breaking
research, and uplifting society through new knowledge, leadership,

and service." [39] British universities, recovering from years of government neglect, have joined the academic arms race. An advertisement for Cardiff University stated that "the aspiration is to become an international research flagship," headed by "a world-class team capable of breaking new ground and producing innovative and high quality science." [40] The University of Manchester aims to be "one of the top 25 universities in the world" by 2015. [41]

Unfortunately for scientific ambitions in the United States, federal agencies normally provide little money for academic laboratory construction, correctly assuming that universities will somehow pull it together from private foundations, state governments, bond issues, and rich alums hankering for their names to be immortalized at the entrance of the new chemistry building. Down to the level of benches on campus quads, few structures in academe are without a nameplate identifying the donor. That the system works all the way up to the mega-level is evidenced by the new laboratories, construction cranes, and rising steel girders on the campuses of research universities throughout the country. Between 1988 and 2001, universities nationwide increased their lab space from 112 to 155 million square feet. [42] Between 2001 and 2003, the total rose to 173 million square feet. At least $7.6 billion was invested in academic laboratory construction in the latter two years, according to the National Science Foundation, which noted that over half of the new research space was designated for biological and medical research. [43] With construction plans stretching to 2020, the University of California, San Francisco, is building a goliath research campus on a forty-three-acre site at Mission Bay, a blighted industrial area south of downtown. Collaboration with biotechnology firms, which are encouraged to build their own facilities nearby, is a prime goal for the new campus. Following a statewide competition, Mission Bay was designated as the site for the management headquarters of California's $3 billion stem cell research program. Cramped for laboratory space, Columbia University plans to build a new campus on an eighteen-acre site in West Harlem, five blocks north of its main campus. In 2006 "slightly more than $200 million" was donated for a research building on the site by the widow of an alumnus and the foundation he established. The University of Michigan, though hard hit by cuts in state appropriations, has a spacious, gleaming, new 240,000-square-foot Life Sciences Institute, built at a cost of $100 million. With hundreds of millions of dollars in state and private funds, the University of Wisconsin–Madison has embarked on a huge expansion of basic science and medical research facilities. A 500,000-square-foot building for stem cell research is included in Harvard's long-term expansion from Cambridge to a 200-acre site across the Charles River in Allston. The federal government is building half a dozen high-containment laboratories specifically for bioterrorism research. The frequency of groundbreaking and ribbon-cutting ceremonies for these and many other new laboratories throughout the country has raised concerns about the wherewithal to staff and run them. Harold Varmus, president and CEO of Memorial Sloan-Kettering Cancer Center, helped lead the campaign for the five-year doubling of the NIH budget while serving there as director from 1993 to 1999. He notes that planning and construction of a major new laboratory takes five or six years, and the first of the new labs are coming on line just as the NIH budget has leveled off.

There's still plenty of money in the system but, perhaps more so than ever, not enough. Since endowment principle is universally considered untouchable, ambitions exceed or stretch financial resources at even the richest universities. In this setting, presidents and their money managers are on alert for new streams of revenue. Quite reasonably, they seek to spin wealth from the knowledge produced in their own laboratories. These activities are commonly embellished with pronouncements about the modern university's public-service responsibilities. The production and application of science and technology for health, job creation, and prosperity are prominently listed among them.

American universities have led the way to market, but universities in many countries, urged on by envy of American technological enterprise, are hurrying to catch up. When we met in 2005, Richard Smith, former editor of the *British Medical Journal,* told me of a recent discussion at London's St. George's Medical School governing council, of which he is a member:

> There is a tremendous emphasis on looking around universities and seeing what there is that is potentially of commercial value, and turning it into a business that will bring income to the university. Places like Oxford and Cambridge have been doing it pretty successfully for quite a while, but some of the other universities are adopting it more slowly as it becomes increasingly obvious to people that they're expected to do more and more with ever-more diminishing funds. So clearly, the money has to come from somewhere else.[44]

Smith observed that making money is a motivation but noted that it's also argued that moneymaking opportunities provide incentives for speeding discoveries to the bedside. He added, "It interested me that there was no discussion about, well, this could lead to all sorts of conflicts of interest, and it could lead to distortion of the research and teaching agenda."

Not long afterward, the *Economist* reported, "British academics are, to an extent that those who work in mere private enterprise cannot imagine, obsessed with money. Partly this is because there is never enough: a huge government-mandated expansion in student numbers since the 1960s has not seen a matching rise in funding. . . . Cross subsidies are common: research grants don't cover overheads and can require matching funds to trigger them."[45]

See You in Court

If getting research from campus laboratories to the bedside or the electronics supermarket were the predominant goal, commercial litigation and strife would be rare among the high performers of academic technology transfer. Why fight over money or glory if the object is to deliver the goods to the public? But battles over patent rights for the income they provide are hardly uncommon in academe. In our knowledge-based economy, intellectual property law is a growing specialty. The bite that legal costs are taking out of the research economy is difficult to ascertain, but a Stanford law professor, John H. Barton, has pointed out—with apparent concern—that "the number of intellectual property lawyers in the United States is growing faster than the amount of research."[46] Since the growth patterns of legal specialization bear some close relationship to the marketplace, we can assume that jousting over patents and other such intangibles is a good business. Science and litigation are a twosome in several notable cases of recent times.

A national leader in income from patents derived from on-campus research—$143 million in 2001—Columbia University balked at the inevitable: the approaching expiration of the patents on its biggest single winner, the method devised by Professor Richard Axel and colleagues for inserting genes into a cell, the so-called cotransformation technique. In 2000, in anticipation of the imminent expiration of three related patents protecting the Axel technique, Columbia enlisted a loyal alumnus, U.S. senator Judd Gregg (R-New Hampshire), to introduce legislation extending the patents for fourteen to eighteen months. Legislative success would produce many more millions in income from the biotechnology firms that licensed the patents so that they could employ the technique for research or in manufacturing their own products. These included Genentech, which reportedly paid Columbia over $70 million in royalties during the life of the patent; Biogen, $35 million, and Genzyme, almost $25 million, according to a report in *Science,* which noted that the total take was estimated at "hundreds of millions of dollars."[47] Feeling they had paid enough, the companies hurried to their own friends on Capitol Hill and blocked the senator's proposal. Whereupon Columbia turned to the U.S. Patent and Trademark Office and applied for and received a new patent for the Axel technique in 2002, claiming novel elements not included in the original patent applications. This maneuver evoked a batch of lawsuits from Columbia's angered biotech customers. Specialists in patent intricacies and strategies credited Columbia with resorting to a "submarine patent," that is, a patent that lies in wait for unsuspecting prey who have unwittingly committed infringements. In 2004 Columbia relented, saying it would not attempt to enforce its newly revealed patent against nine of the suing companies; in the following months, it settled with all but one of the other companies, in all instances on confidential terms.[48] Outsiders were thus denied a look at a messy and costly underside of commercialized academic science. "All big winners end up in litigation," I was told by a veteran of courtroom wars arising from academic patenting, Scot G. Hamilton, senior director and patent attorney in Columbia University's Science and Technology Ventures.[49]

Columbia has not revealed the costs of its failed patent foray in Congress or its unsuccessful attempt to breathe new life into its expiring money-spinners. But the expenditure of legal fees "in the low millions" was acknowledged in 2004 by the general counsel of the University of Rochester in its unsuccessful litigation to win billions of dollars from Pfizer, Inc., and Pharmacia, manufacturers of best-selling painkillers based on the COX-2 inhibitors, which Rochester credited to its own laboratories. A ruling against Rochester by a U.S. District Court judge was upheld by the U.S. Court of Appeals for the Federal Circuit; the U.S. Supreme Court declined to take the case. Loyally coming to the assistance of a university seeking financial rewards from science, eleven other universities submitted briefs in support of Rochester, though several others declined to join the struggle.[50] Rochester's willingness to gamble scarce millions on a patent suit reflected a hunger for money that pervades academe. In the mid-1990s, Rochester was so hard-pressed for money that it announced a "Rochester Renaissance Plan," which mysteriously included elimination of its graduate program in mathematics. Rochester ultimately backed off under withering denunciations from mathematicians and other academics around the country.[51]

"Litigation is a sign that you're doing some things right, because nobody is going to litigate over nothing," Lou Berneman, then managing director of the University of Pennsylvania Center for Technology Transfer, told me. "It makes sense that people are taking shots at the Columbia cotransformation patents. If you're doing things right, then there's some value involved, and there are going to be people who want to see the other side of that. Money is the mother's milk of science," he assured me.[52]

And so it was in a case literally involving milk and money, settled for at least $185 million, in an eve-of-trial agreement in 2006 between the University of California and Monsanto Company. At issue was the patent for a hormone for boosting milk production, bovine somatotropin. Citing scientific discoveries at its San Francisco campus, the University of California sued Monsanto in 2004 for infringing the patent for the product, marketed under the name POSILAC. In the settlement, Monsanto agreed to pay the university a $100 million upfront royalty, plus fifteen cents per dose of POSILAC sold to dairy producers, with an annual minimum royalty of $5 million through the 2023 expiration of the patent.[53]

Donald Kennedy was president of Stanford University from 1980 to 1992, a period in which Stanford and nearby Silicon Valley fabulously enriched each other through exchanges of innovative talent and money in an intensely entrepreneurial atmosphere. Kennedy recalled:

> Everyplace I went for a while during my time as president of Stanford, people would ask me, "How can we become the new Silicon Valley?" They would ask that in "Vulture Canyon" someplace. These relationships and the opportunity to derive significant revenue from transferred technology are scarce, and they depend a lot on local circumstances and relationships. And I'm not surprised that most of the people who have ginned up a whole new apparatus to make money out of this aren't doing it.[54]

Until it reaped a bonanza from its investment in Google, the top moneymaking patent for Stanford was the Cohen-Boyer patent for gene splicing. The technique, developed in collaboration with the University of California, San Francisco, provided the scientific foundation for the multibillion-dollar biotechnology industry. "You know what the second leading patent for Stanford was after the Cohen-Boyer patent?" Kennedy asked. "It came from the music department, and it's the chip that goes in Yamaha synthesizers. Who would have thought?"

Unrestrained chasing of commercial dollars is considered indiscreet in academic institutions. And, as we've seen, many researchers are uninterested in commercial activities. But the moneymaking potential of campus laboratories is so widely recognized throughout higher education that it's difficult for an academic scientist today to be unaware of the opportunities that may be out there, or for a graduate student to fail to be aware of commercial interest mixing with science. It is also challenging for managers of the system to navigate their way to new sources of money while avoiding snares and taints that are common to commercial activities but are considered unseemly or incompatible with the canons of science. Sloan-Kettering's Harold Varmus summed up the dilemma to me as follows:

> In general, my own board of trustees is very concerned about getting the institution in trouble about conflict of interest, something that backfires, bad news in the *New York Times*. All these things have happened in the past in one guise or another.[*] They'd be very detrimental to the institution. And they're also quite wary of the idea that we've become unhooked from our primary motive, which is to understand more about cancer cells and make basic discoveries. That's always been a major motivation here. And if we become tainted by the appearance of just becoming a drug company, that would be very upsetting to most of my trustees. On the other hand, especially in this period of intense growth—we have a big building going up here, we're doing a lot of things in the clinic and elsewhere, a new graduate school. All kinds of things that are expensive. And nobody wants to miss an opportunity, like the opportunity we had with G-CSF [granulocyte colony-stimulating factor, a drug, licensed to industry, producing hundreds of millions of

---

*In 1974, long before Varmus was appointed, the Sloan-Kettering Institute for Cancer Research was the scene of a highly publicized case of scientific misconduct arising from published reports of a new technique for transplanting unrelated tissues without rejection, thus overcoming—if true—a major impediment to organ transplants. The reports were substantiated by the presence of black patches on a white mouse, ostensibly evidence of successful transplantation. Alas, the black patches came from a felt-tip pen wielded by the perpetrator, William T. Summerlin, a promising young researcher at the institute. The episode is described in a standard work on scientific fakery, William Broad and Nicholas Wade's *Betrayers of the Truth: Fraud and Deceit in the Halls of Science* (New York: Simon & Schuster, 1982), pp. 153–57.

dollars in royalties for Sloan-Kettering]. On the other hand, there's considerable resistance to the idea that we're going to overprotect [by not sharing scientific information and by patenting discoveries].[55]

The pursuit of money is at the heart of modern university administration. The presidents are judged by their fund-raising prowess. Rare among them today is a statesman or philosopher of higher education. If they moderated the growth aspirations of their institutions, they could get along with less money. But that would amount to capitulation in the academic arms race—a sure signal for change of command. The quest for untapped sources of money is the preoccupation of the chiefs and their staffs.

2 Elusive Industrial Angels

And as federal funds get tight, and institutions have built up capacity during the heady days of the doubling of the NIH budget, I think it's inevitable that they're going to turn more and more to industry to amortize their investments in people and buildings. I don't think that industry funds anything that isn't part of business. It may be very near term or it may be a little longer term, but they're not philanthropic.

David Korn, senior vice president, Association of American Medical Colleges; former dean, Stanford University School of Medicine[1]

: : :

The federal government's spending for research rapidly expanded following World War II, but never enough to pay for all the ambitions and potential in science. For good reasons, industry came to be regarded as a promising source of additional revenue for science in universities. Big companies, many prospering on new technologies, possessed the only big pot of money that universities weren't already tapping for significant amounts. Collaboration between the two sectors seemed natural and mutually beneficial. For several decades the industrial support did increase, thus encouraging academic hopes for continued growth. Alas, today, contrary to many misunderstandings, industry is a very minor participant in the financing of academic research. Meanwhile, sources other than industry—federal agencies, endowment income, philanthropic foundations, tuition, private gifts, state appropriations—provide virtually all the money for higher education and its financially ravenous scientists, though never enough. Corporate America is parsimonious in its financial support of academic research, despite the widely held belief that business and industry financially dominate university science and medical research budgets and thereby call the tune.

In 2005 universities spent $45.8 billion on R&D. Of that amount, the federal government provided $29.2 billion. Industry provided $2.3 billion. Over the past quarter century, industry's share of academic R&D financing hit a peak of 7.4 percent in 1999 and a low of 4.9 percent in 2004, according to the National Science Foundation.[2] Within academe, however, the dreams of industrial money live on, because the other sources are stretched and industry does indeed spend a great deal of money on research—far more than the U.S. government, again contrary to general misunderstanding. But very little of industry's R&D money is spent in universities. Rather, the industrial money stays close to home, in corporate America's own laboratories and, increasingly, in company-owned laboratories abroad because they are cheaper to operate and closer to local markets.

In 2005 industrial firms budgeted $191 billion of their own money for research and development, and spent another $20 billion or so in R&D contracts from federal agencies. In the totality of national spending on R&D, industrial firms exceeded the federal government by about two to one.[3] The proportions are obscured because industry's money is distributed among many firms, whose research programs and budgets are a relatively small and little-publicized part of their overall corporate operations. For competitive purposes, many companies keep quiet about their research. Most government science programs, in contrast, are publicly announced, scrutinized, and squabbled over by Congress and reported in the press, thus generating a buzz disproportionate to the amount of money involved. The leading exception to this pattern is the pharmaceutical industry, which publicly boasts of its research expenditures to justify exorbitant prices for new drugs.

Among university scientists and administrators, there's a righteous conviction that industry owes academe a great deal. The sentiment is based in fact, though without effect on industry's tightfisted preference for avoiding expenditures, if it can. Universities train the scientists who create wealth in business and industry. That's where one-third of all science and engineering PhD's are employed, a figure that nearly doubles when master's degree holders are included.[4] Moreover, research in academic laboratories produces intellectual feedstock for high-tech industrial products and services. By all rights, it seems plain and just, at least to academics, that industry should bear a substantial financial responsibility for the university-based science that greatly contributes to corporate success. However, contrary to popular impressions, and academic hopes and expectations, industrial firms and other profit-seeking enterprises provide a surprisingly small share of university research budgets.

The NSF's tabulations show this is true among the small fry as well as the mighty powerhouses of academic R&D. A representative handful from data compiled by the NSF shows that Harvard, with $454 million in R&D expenditures in 2004 (mostly from government agencies, as is the case with all universities) reported $5.8 million from industry. R&D totals and industry shares for several other universities: University of Washington, $714 million/$46.5 million; Yale University, $423 million/$14 million; Vanderbilt University, $318 million/$5 million; Boston University, $240 million/$8.5 million; Brown University $130 million/$1.7 million.[5]

The paucity of corporate research money for academe was highlighted in 2006 when Bristol-Myers Squibb (BMS) canceled its Freedom to Discover program, which since 1977 had provided no-strings funds for university-based scientists to conduct biomedical research of their own choice. Costing about $6 million a year at the time of the cancellation, the program provided some fifty scientists $100,000 a year for five years. According to *Science*, BMS described the program as "the largest corporate-funded, unrestricted research grants program in the world."[6] The cancellation was attributed to a corporate decision to shift the funds to pediatric AIDS clinics in Africa and concerns that the "company could run afoul of new, restrictive regulations in Europe on corporate gifts to physicians." For 2005 BMS reported profits of $3 billion, a 26 percent increase, on sales of $20.2 billion.

Industry is not unaware of its self-interest in a productive academic research enterprise. In the boardrooms and executive suites of industrial firms, there's a clear understanding of the economic importance of academic research. When Congress threatens to skimp on taxpayer-financed research, as in 1995 when the newly elected Republican majorities went on a budget-cutting spree, corporate America protests—on informed, selfish grounds. On that occasion, the CEOs and former CEOs of sixteen Fortune 500 high-tech firms warned in a full-page advertisement in the *Washington Post* that government spending for science in universities was crucial to their business success. "We can personally attest," they stated, "that large and small companies in America, established and entrepreneurial, all depend on two products of our research universities: new technologies and well-educated scientists and engineers."[7] Nothing prevents corporations from emulating the federal government's provision of a great deal of money for research in universities. Industry, however, prefers to have the taxpayers foot the bills. Companies do sponsor research in universities when it clearly benefits their business operations, but most academic research is too far "upstream" for quick commercial exploitation. Moreover, unless patentable, which is not often the case, the science produced by professors is openly published as quickly as possible. Why pay for it when it's available for the price of a scientific journal? Of all the money that industry spent on research in 2004, only 1.1 percent was spent in universities, a drop from 1.5 percent a decade earlier.[8]

The uninitiated in academic finance should understand that the industrial share of dollars for universities is even smaller than the NSF numbers suggest. R&D is only one of many activities in a university budget—a big one, but it usually takes up no more than a quarter to one-third of annual expenditures in a major research university. As a share of *total* spending in these universities, the industrial R&D money is a small, often tiny, part of the budget. At the Massachusetts Institute of Technology, unabashedly entrepreneurial and as deeply entwined with industry as any university, industry provided $81 million of $486 million for R&D in 2003, while federal agencies provided $356 million.[9] In that year, the campus-wide budget for all MIT activities, science-related and otherwise, was about $1.4 billion. Industrial money thus amounted to 16.7 percent of the R&D budget and 5.8 percent of the total budget. Net income from MIT's patented discoveries licensed to industry was $15 million, according to Lita Nelsen, director of MIT's Technology Licensing Office. Referring to the licensing income, she laughingly observed, "On a billion-point-four budget, it's not going to change the price of tuition at MIT."[10] In later years the amount rose, but the proportions remained about the same.

Some quibbling about the volume of corporate money in academe is justified but doesn't change the basic point: industry provides a relatively small portion of the money that supports R&D in universities. The NSF, which tracks research money, is in an unsettled state of mind about accounting for the billions of dollars that industry spends on clinical trials in academic health centers. That money often moves along nationwide or, sometimes, worldwide routes, managed at a university that was selected to organize and coordinate a trial, but spent at near and distant medical centers and hospitals where the drugs are tested on volunteer patients. Some major research universities, including MIT, do not have medical centers and conduct little or no clinical research. Georgia Tech, lacking a medical center, teams with neighboring Emory University to participate in clinical research. But at universities that are equipped to test drugs on humans, the clinical-trial receipts are substantial, and sometimes the research itself becomes ethically troublesome. It is in university-based clinical drug trials financed by pharmaceutical firms that some of the more infamous dirty deeds have occurred, such as suppression or misrepresentation of findings unfavorable to the corporate sponsors, failure to give patients adequate information about the risks of experimental treatments, and testing and evaluation of drugs by clinicians with financial interests in the outcome. Some universities report their clinical-research sums to the NSF's R&D scorekeepers; others don't. In 2006 the NSF was preparing to study the matter.

The NSF's tabulations of industry money on campus focus on research that outside corporations finance inside the university. It omits several other types of income that entangle scientists and business. These include equity in university-related start-up companies and professorial income from industrial consulting, in total unknown but substantial for some individuals. So, in one way or another, industry's financial presence in academic science is unquestionably bigger than the reported amounts. Nonetheless, it is far outdistanced by the major source of money for science in universities, the federal government. The official statistics of R&D are voluminous and complex, embodying varying definitions of types of research, gaps in the data, and cautionary footnotes. But in its 2006 biennial report on the state of R&D, the National Science Board summed it up: "Industrial support counts for the smallest share of academic R&D funding, and support of academia has never been a major component of industry-funded R&D." The report noted that in 1999 industrial support of R&D in universities peaked at 7.4 percent. By 2003 it had declined to 5 percent.[11] Since then it has changed very little.

Slippery Stats

Statistics can mislead, and when they are meshed with intimations of unwholesomeness or worse, beware—and even more so when an echo effect is present. Let's look at one example, starting with an assertion, published in 2003, by a leading critic of academic-industrial business relations: "The massive infusion of private R&D is changing the character of some institutions," Sheldon Krimsky wrote in *Science in the Private Interest: Has the Lure of Profits Corrupted Biomedical Research?* "By 2000," he stated, "Duke University had 31 percent of its [R&D] budget funded by industry."[12] A favorable review of Krimsky's book in 2004 in *American Scientist*—journal of Sigma Xi, a national scientific society—said, "Commercial funding today is the lifeblood of biomedical research, Krimsky observes."[13] In 2005, citing Krimsky in her book *University, Inc.: The Corporate Corruption of American Higher Education,* Jennifer Washburn wrote: "Duke University now draws 31 percent of its research and development budget from industry." Reporting other universities with double-digit industrial R&D funds, Washburn added: "What's truly new—and dangerous—is the degree to which market forces have penetrated into the heart of academia itself, causing American universities to look and behave more and more like for-profit commercial enterprises."[14] These numbers and assertions warrant scrutiny if we are to plumb the polemics and realities of science for sale.

Examination of data provided by Duke University upon request confirms that the 31 percent figure for the year 2000 is essentially correct but also possibly misleading, deserving of exploration, and, with the passage of time, out of date. In 2000 Duke's campus-wide budget totaled $2.1 billion. Within that sum, R&D expenditures totaled $356.6 million, of which industry provided $109.8 million. Thus, industrial money amounted to 30.8 percent of Duke's R&D spending, but only 5.2 percent of Duke's total spending in 2000. A closer look reveals that over half of the $109.8 million that industry provided for R&D was spent at the Duke Clinical Research Institute (DCRI), part of the Duke University Medical Center. The DCRI is described by Duke as "the world's largest academic clinical research organization." Its nine hundred faculty and staff members orchestrate drug trials involving over five thousand researchers and hundreds of thousands of patients in scores of countries around the world. The DCRI's "unusual size and scope," Duke stated, "tend to skew the figures."[15]

Whether the cited 31 percent that industry provided for R&D at Duke is excessive or not is a matter of accounting and ideological taste. But it is worth noting that clinical research—to assess the safety and therapeutic value of drugs and other treatments—is deemed critically important by government and academic health-care strategists. Duke has established itself as a world-leading center for clinical research. More clinical research is needed, we are often told. The NIH "roadmap" for clinical research in the twenty-first century states that the "translation" of research into treatment "lies at the very heart of NIH's mission," and cites the "need to develop new partnerships of research with organized patient communities, community-based health care providers, and academic researchers." [16] In recent years the NIH has greatly expanded its own facilities for clinical research and has increased funds for clinical trials and training for clinical researchers at universities. In 2005 the NIH announced that $11.5 million in Clinical and Translational Science awards would be made available for planning up to fifty new centers and that by 2012, $500 million would be provided for funding their research programs. The large role of academic medical centers in clinical research is neither surprising nor undesirable.

Bringing Duke and R&D closer to date, by 2003 the budget for everything at Duke totaled $2.8 billion. R&D funding from all sources, particularly the federal government, had rapidly grown over the preceding four years. Within the overall budget, R&D now totaled $520 million, of which $122 million was provided by industry—with "well over half," according to Duke, spent on clinical research. Thus, the corporate cash that year amounted to 23.5 percent of the research budget and 4.4 percent of the university-wide budget—a figure that does not suggest "a massive infusion of private R&D." [17]

Business on Many Fronts

Academe's involvement with commercialization is not confined to company-sponsored research and related activities in university medical facilities and laboratories. Universities also go off campus in quest of commercial winnings, by licensing their discoveries to industrial firms in return for fees and royalties and by assisting in the creation of start-up companies to develop products from research initially conducted on campus. All these activities benefit from and intensify the entrepreneurial spirits on campus. On many fronts, academic science and commercialization find points of contact and collaboration. As might be expected, these relationships are not invariably successful, and on some occasions, whether successful or not, they collide with traditional academic values and produce an uproar that in turn affects the environment for further academic-corporate dealings. The customs, rules, and expectations surrounding the relationship between the two sectors are not static. Driven by scandal, episodes of outrageous behavior by academics, and ideological piety, regulations governing their dealings

have evolved in recent years, mainly in the direction of required disclosure of professorial financial connections with industry, stricter limits
on profit-making opportunities for academic participants, and tighter
controls over what companies can extract from academic partners. The
changes emanate from government agencies that finance and regulate
research, professional associations, and from individual universities.
Many academics and administrators regard the new regime as onerous
and inimical to easy transfer of knowledge from campus to corporation. But their squawking is disproportionate to the reality. A good
deal of business continually goes on between universities and corporate organizations interested in their skills, scientific discoveries, and
prestige. There's no truth, however, to the frequent, wholesale depic-
tions of university-based science as a passive appendage of corporate
America.

A Big Deal in Berkeley

In recent years, as we've seen, industry's support of overall academic
R&D has remained small and relatively stable. Meanwhile, the federal
government's share has increased. The reason for lack of growth in industrial support is not clear. But a contributing factor may be corporate
wariness of academe's growing sensitivities and regulations regarding
industrial money. One episode in particular has acquired legendary
status in the folklore of the scientific sellout.

An ideological blowup was assured in 1998 when the Novartis Agricultural Discovery Institute, a Swiss pharmaceutical firm,
bought privileged access to research in the Department of Plant and
Microbial Biology at the University of California, Berkeley (UCB), for
$25 million spread over five years. (Novartis was later affiliated with
and then separated from Syngenta, which inherited the agreement with
Berkeley.) This was not the first big-dollar industry-academic deal of
its kind. In 1982 Monsanto entered into a long-term collaboration
with Washington University in St. Louis. Harvard and other universities also took on major solo funding from corporate sponsors. But the
Novartis deal, as it is usually referred to, was among the few major
corporate couplings with a state university—in this case, California's
politically volatile Berkeley campus, home of the free speech movement
of the mid-1960s and ever since a major site of antiestablishment spirits. The lure of money overcame doubts within the university about the
propriety of selling science in a public institution to a single corporate
buyer, even with an assortment of presumed safeguards for academic

values and the public interest written into the contract. The outcome, however, was not as intended by either party: Berkeley got a black eye and Syngenta received virtually nothing of unique scientific value for the $25 million.

Apart from raising both departmental salaries and widespread indignation, the deal proved to be a dud in virtually all respects, according to a 188-page independent inquest report financed by the University of California Regents and performed by researchers from the Michigan State University Institute for Food and Agricultural Standards. "Few or no benefits, in terms of patent rights or income, to either UCB or Novartis/Syngenta have emerged from research conducted in the course of the agreement," the report concluded. It added, "The direct impacts of UCB-N [Novartis] on the university as a whole have been minimal. The agreement has not produced the major changes that many feared it would." A positive effect cited by the reviewers was a doubling of the graduate program, in connection with which it noted that "among post-doctoral researchers, salary was deemed to be the greatest benefit of UCB-N." [18] An initial denial of tenure to an academically well-qualified outspoken opponent of the Novartis deal created a secondary explosion, followed by a reversal and the granting of tenure. The most long-lasting effects of the Berkeley-Novartis deal were widespread condemnations of departmental buy-ups, or sellouts, as unholy, plus doubt about the substantive value of such blockbuster deals for both university and corporation. The conscience of academic purity, the American Association of University Professors, weighed in with an admonitory pronouncement on "Corporate Funding of Academic Research," warning that no amount of legal ingenuity could immunize a deal of that kind against malign effects:

> Where the financial resources of an academic department are dominated by a corporation, there is the potential, no matter how elaborate the safeguards for respecting academic freedom and the independence of researchers, for weakening peer review both in research and in promotion and tenure decisions, for distorting the priorities of undergraduate and graduate education, and for compromising scientific openness.[19]

Assessing the deal pragmatically, former Stanford president Donald Kennedy told me he was doubtful about the effectiveness and future of blockbuster corporate funding of academic research: "Big deals between big companies and universities have not been successful, by and large," he said. "And so I think people are just deciding that they don't work and they're complicated and hard to bring off, so why try 'em?" Kennedy said:

> We did well [at Stanford] with a Center for Integrated Systems that a lot of electronics companies co-participated in. We had a helluva problem dealing with intellectual property interests until the eight or ten participating industries agreed that nobody would take a stake. And it just sort of became an opportunity for those companies to send people to visit, to collaborate, to look over graduate students. And it was a good opportunity in that case for both parties, because nobody took anything out the door. Universities really need to remind themselves that public regard is the best thing they've got going for them. And friends in Congress and in the polity are the best friends they have.* Their traditional reputation is of being very socially dedicated and not driven primarily by profit. And to the extent they impinge on that or deprecate it, they're losing.[20]

Nonetheless, though less frequent today, big corporate funding retains a presence in academic research. In 2002 Stanford University launched a Global Climate and Energy Project, priced at $225 million over ten years. ExxonMobil, a declared disbeliever in global climate change and generous angel for right-wing think tanks, was chief sponsor and donor of up to $100 million for the project. Other corporate contributors in the venture included General Electric, Toyota, and Schlumberger, another energy-related firm. Snide commentary about big corporate money in academic research soon followed. The *Chronicle of Higher Education* reported that some environmentalists and academics saw the deal as an attempt by ExxonMobil to "'greenwash' its environmental reputation" for a price akin to its "paper-clip budget."[21] Stanford officials were prompt with assurances that the project would respect traditional scientific and academic values. Among them was Stanford professor Franklin M. Orr Jr., director of the project, who said that "there are plenty of protections built into the agreement

---

*On the topic of popular and political regard for universities, Kennedy is a battle-battered witness, having been pilloried in Congress and in the press for allegedly misusing government research funds for personal purposes while president of Stanford. The allegations, which forced his resignation, are candidly discussed in Kennedy's book *Academic Duty* (Cambridge, MA: Harvard University Press, 1997).

that ensure there's plenty of independence for our researchers."[22] In addition to providing for university employees to serve as director and deputy director of the project, the contract called for Stanford to establish an advisory committee composed of members unaffiliated with the university or the industrial sponsors. Their role, the contract specified, "will be to provide constructive, informed, broad-based advice to the University and Sponsors on the content, direction, quality, and progress of the Project."[23] The possibilities for bias favorable to the corporate sponsors cannot be discounted, but in the current era of heightened sensitivity to abuse of academic integrity, the risk of public opprobrium for offending accepted values is substantial.

Apart from the possible dangers to academic integrity, both real and assumed, a separate issue is present: Why did one of the world's richest universities even risk the suspicion of commercial contamination in a project that is politically sensitive, technically important, and well within its means? With an endowment of over $12 billion and annual gifts and government research support adding another billion, Stanford could easily finance the project from its own ample resources. But universities are always hawk-eyed and eager for outside money. No amount that they have in the bank can keep them from pursuing yet another donor, even one with the ideological baggage of ExxonMobil. Though seared by the Novartis deal, Berkeley entered into another mega-industrial deal in 2007, agreeing to lead a $500 million energy-research consortium financed by BP, formerly British Petroleum. The customary ideological howls were raised, but to no avail.

The big difference today is public and academic scrutiny and the specter of embarrassment or disgrace for ethical shortcomings. Together they produce strong, perhaps irresistible, academic insistence on shared governance over use of industrial money; quick, if not immediate, publication of the results; and adherence to academe's concept of the rules of the game. We must not naively assume that academe of its own volition has ascended to a higher level of integrity. It *has* ascended but mainly because the risks of getting caught and hurt for bad deeds have greatly increased.

Given the skeptical scrutiny that now accompanies big money deals between universities and corporations, it's to be expected that many business executives justifiably wonder whether they're worth the risk of public-relations bruises. Needy as most universities are, many would be delighted to sell a chunk of themselves to a corporate sponsor, preferably on conditions that respect their chastity, though that's negotiable at some schools. But not many corporations are besieging universities to take their money. Eagerness for even more business is strongest on the university side of the relationship—ever famished for money. Contrary to polemical assertions and popular belief, industry tends to be cautious about entanglements with academe, put off by its mounting rules for protecting intellectual property rights and scientific integrity. Citing conflicts that can arise over patent and licensing rights for industry-financed research in universities, Susan Butts, a Dow Chemical Company executive, told *Nature:* "Fewer and fewer companies want to work with universities on sponsored research, because they feel it doesn't make good business sense. Companies could disadvantage themselves if it produces inventions that they are ultimately unable to license."[24]

Another perspective was offered to me by Karl Koster, MIT's director of corporate relations:

> Companies are in the business of making money. Some of them are very secretive about what they do. And there are companies that like to contract for research, and there are various mechanisms outside the university system. Some companies really don't understand that the real benefit of working with MIT is really their investment in their human capital. Their brightest technical people can understand the leading edges of research in domains that are important to the company. That's the principal benefit of getting involved with a university. It's really an investment in their own knowledge acquisition and capital. Recently there's been a lot of press about intellectual property, but if you look at MIT, we had about $570 million worth of research on campus last year. We applied for two hundred and something patents. Every $2 million, you get maybe one patent. And how many of those patents actually wind up earning money or being a product or some subset of that? If all a company is after is acquiring intellectual property, I would say there's a lot cheaper ways to do it than getting involved with a university. We don't do proprietary research. If what they're really after is acquiring intellectual property, I think they're much better off acquiring it with companies that have it, not by working with MIT.[25]

It may be assumed that if industrial firms could make more money by spending more money in universities, they would do so. The failure to spend more overall should invite wonder, perhaps skepticism, about the depiction of academic scientists as prize prey for corporate exploitation. Maybe they're complaisantly giving it away to corporate seducers. But that would be out of character in contemporary science.

In any event, intimacy between academic science and corporate America is not optional. It is required by an act of Congress.

3 Commercialize! It's the Law

It is the policy and objective of the Congress to use the patent system to promote the utilization of inventions arising from federally supported research or development . . . [and] to promote collaboration between commercial concerns and nonprofit organizations, including universities.

Bayh-Dole Act, 1980

: : :

Starting in the mid-1960s, as foreign manufacturers prospered in American markets, public and political attention was drawn to the steady increase of government expenditures for research in universities, from $435 million in 1960 to $2.5 billion in 1975.[1] At that pace of growth and financial level, science spending became conspicuous and politically interesting. Scientific inquiry as a manifestation of the human spirit is an inspiring notion, but politicians wanted tangible results, not just arcane research papers. What were we getting for the money? Scientists and their friends defended the spending and called for more, with assurances that science, both directly and in its own serendipitous fashion, would produce excellent economic returns, cures, weapons, and other benefits. Though economists might fret about opportunity costs and cost-benefit ratios, the science promoters, indifferent to economic theory, confidently insisted that good alone resulted from expanding government spending for research and training of more researchers. But a serious traffic problem was reported on the road from government-financed university research to the marketplace. From many quarters came the contention that valuable patents based on government-financed scientific discoveries in universities remained outside the economic mainstream because, by law, they belonged to the federal government or the ownership criteria were ambiguous. Law, or the lack of it, was thus seen as blocking the advance of an important economic process, technology transfer, from campus to corporation. Prior to 1980, the federal government owned thirty thousand patents, of which only about 5 percent were licensed to industry, according to one of the patriarchs of academic tech transfer, Howard Bremer, former patent counsel at the Wisconsin Alumni Research Foundation.[2]

Among the research universities, attitudes varied about the propriety of seeking patents and industrial customers for the discoveries of their scientists. Some universities had long before lost inhibitions about profiting from science. For universities that desired patent protection for their research but preferred to remain distant from commercial dealings, assistance was available from the Research Corporation, a university-oriented, nonprofit patent service founded in 1912. Tech transfer is distinct from scientific research and has never been an easy process. Matching academic science and industry is specialized work, requiring knowledge of the needs and interests of industrial firms, evolving markets, and the commercial potential in patented concepts that are far distant from saleable products. A precursor of the tech-transfer offices now found in every research university, the Research Corporation located industrial customers for academic patents, negotiated deals that produced revenues for inventors and their universities, and put its share of profits back into academic research.

Many universities, however, remained queasy about selling their science and shunned opportunities for patents. But pressures for tech transfer were increasing. In 1968 the U.S. Department of Health, Education, and Welfare (HEW), parent agency of the National Institutes of Health, set up a system of Institutional Patent Agreements that enabled universities to take title to their HEW-financed inventions if they were staffed to negotiate their transfer from campus to industry. Similar methods were adopted by the Department of Defense, the National Science Foundation, and other government agencies that financed academic research. However, the transfer systems remained a patchwork of rules, based on individual agency policies rather than federal law applicable government wide. Twenty-six agencies played according to their own patent rules, Bremer complained, and the muddle was further compounded, he said, when several agencies joined in financing a project. In that case, "the most restrictive agency policy became the controlling policy."[3] Under these fragmented arrangements, the rules were subject to conflicting interpretations and abrupt, politically induced changes, which caused industry to be wary of patent deals with universities if government research money was involved.

Giveaway or Competitive Handicap?

Lingering in the political background was long-standing ideological contention over the ownership of government-financed research. After World War II, legislation to create the National Science Foundation was stalled for five years while congressional factions and the White House fought over control of the new bankroll for science and the issue of ownership of patents arising from NSF grants. Populists invoked the specter of giveaways, warning that corporations would scoop up patents based on government-financed university research and use their economic and political power and wiles to dominate markets and throttle innovations that threatened their products. Free-enterprisers contended that antiquated rules and restraints on profit-making bottled up academic science and handicapped American industry. Some industrialists demanded the right to exclusive licenses for developing products from university patents. In their view, exclusivity was crucial, because without it, industrial firms were prudently reluctant to undertake the costly work required to transform scientific findings into saleable products. Why invest in developing a product if a competitor might also license the patent and divide the market or beat you to market, perhaps with a superior product? In fact, nonexclusive licensing was common and was comfortably accepted by many companies, more so for patented research techniques that scientists employed in industrial and university laboratories, but also for patents on manufactured products for the general public. Nonetheless, legislation allowing exclusive licensing of patents arising from government-funded research was strongly advocated as necessary to revitalize American industry.

In bipartisan harmony, Senators Bob Dole (R-Kansas) and Birch Bayh (D-Indiana) introduced a legislative prescription that easily won congressional passage. Senator Russell Long (D-Louisiana), son of "Every Man a King" Huey Long, put up a lonely fight, denouncing their legislation as "one of the most radical and far-reaching giveaways I have seen in the many years I have served in the United States Senate."[4]

The Bayh-Dole Act (35 USC §200–212), a wordy amendment to the Patent and Trademark law, was passed in 1980, with scant interest on Capitol Hill and in the press, both preoccupied that year with presidential and congressional elections and the captivity of the American Embassy staff in Teheran. Carter lost his reelection campaign and signed the bill as a lame-duck president. Bayh, too, was defeated for reelection and became a Washington lobbyist. However, in the policy circles that encompassed government research officials and university administrators, a victory was hailed. Swept away was the scattered collection of regulations governing the ownership and sale of government-financed research in universities. Taking its place was a federal law that not only gave universities clear title to the patents, but also imposed on them and their scientists a duty to pursue licensing to industry as a condition of accepting government money for research. For the nation's research universities, going to market was no longer optional or voluntary. It was their legal obligation. They could escape that obligation only by engaging in the unthinkable: declining government research grants.

In a gesture to the taxpayers, Bayh-Dole provided the government with a royalty-free license to the patents, which has turned out to be a very rarely exercised option. Among other provisions, the bill directed universities to conduct their patent dealings with small business firms, a dual sop to populist sentiments and the presence of small businesses in every congressional district. Three years later, the small-business preference was removed from the act, in recognition of the fact that conventional small businesses lack the talent and resources to transform scientific knowledge into a saleable product. However, involvement for small business later developed on its own with the proliferation of biotech spin-offs from academic research. Under Bayh-Dole, universities owned the patents arising from discoveries by their government-financed scientists. But the same scientists could establish spin-off companies, often in nearby university-owned research parks, license the patents, and gain riches by turning their discoveries into tangible goods; or, more likely, sell their fledgling firm to one of the many big companies that preferred to buy promising research rather than gamble on making their own discoveries. Venture capital from friends and kin, and wealthy individuals (so-called angel investors, best known for staking Broadway shows), university treasuries, private firms, and public agencies nourished the new spin-off economy. In many regions, assistance became plentiful for the entrepreneurial academic.

The Bayh-Dole Act legitimized and compelled private dealing be-tween corporate America and government-financed science in univer-sities. As a law of the land, it absolved academe of its ancient qualms about engaging in commerce, though some universities, Harvard among them, moved cautiously toward the marketplace. At their dis-cretion, universities could license a patent exclusively to a single firm, which would then have the sole right to develop and sell a product, or they could deal with multiple licensees. University laboratories con-duct research. They don't manufacture pills or devices, nor do they normally engage in product research and development. That's what in-dustry does. With the passage of Bayh-Dole, universities and industry now had a clear shot at jointly making money from booming federal expenditures in academic science—if a product is developed and finds sales. As it turned out, this sometimes happened, with spectacular fi-nancial results, but not often.

The legislation legalized and encouraged the process, but it did not ensure that science would move smoothly from campus to corporation and the marketplace. There was still a human dimension to be dealt with. Under the Bayh-Dole Act, universities must disclose to the federal agency that paid for their research any "invention" resulting from the research, with "invention" defined as a discovery that may be patent-able. Publication in a journal or discussion at a meeting can jeopardize access to patenting because of the statutory requirement that an inven-tion must be novel to be patentable. As Stanford University's Office of Technology Licensing explains to the university's scientists, "To satisfy the novelty requirement, an invention must not have been known to the public. . . . Once an invention has been presented to the public, for example, through publication for more than a year, it is no longer considered new in the U.S."[5] For obtaining foreign patents, the novelty rule is even more restrictive, allowing for little or no prior exposure.

Irrespective of their interest in the commercialization of research, scientists who received government grants were thus beckoned into the process. The disclosure requirement, however, was not accompanied by an enforcement mechanism, and it also faced a difficulty rooted in the culture of academic science: with rare exceptions, promotion and tenure in the major leagues of science inflexibly depend on publica-tions, preferably in selective, peer-reviewed scientific journals. Patents, no matter how much genius they embody or how much money they make, traditionally have not been part of the tenure system, or at most have been only vaguely recognized. Efforts to bring patents into tenure

evaluation are rarely well received, as related to me by Michael Douglas, vice chancellor and head of tech transfer at Washington University, in St. Louis. Douglas said he favors inclusion of patents in tenure
evaluations, "but some department chairs will puke on it when you say
that." In engineering and other applied fields, there's more tolerance,
Douglas said, "but sitting up here at Washington University's medical
school, the great tradition that it has for being quite frankly one of the
most academic institutions I have ever seen, I don't think you're going
to get department chairs giving much weight to a patent application."[6]

That may be so, but an exception adopted by a major university
system demonstrates commercialism's power to chip away at the sacred
pillars of academic culture. In 2006 the Board of Regents of the Texas
A&M University system announced that "patents and the commercialization of research, where applicable," would be added to the traditional criteria for tenure. "State funding for education is declining as a
percentage of total funding," Chancellor Robert D. McTeer explained.
"We must recognize that we must rely more on partnerships with business and industry for funding."[7] Commenting on the Texas decision,
Roger W. Bowen, head of the American Association of University Professors, stated that it "reflects the furthering of commercialization of
higher education," adding, "As far as I know, this is the first time such
action has been taken."[8]

In the day-to-day workings of Bayh-Dole, the crucial decision maker
for disclosure is the scientist en route to publishing a paper or deliver-
ing a lecture reporting research findings with commercial implications.
In their eagerness to claim scientific credit and receive recognition for
their work, scientists may unwittingly, or deliberately, squander patent eligibility by telling too much. The tech-transfer specialists who
understand the intricacies of patenting and disclosure cannot possibly follow the progress of research and publishing in the hundreds of
laboratories and research groups on a major university campus. Apart
from what they learn from the grapevine or through personal relations with the researchers, they must rely on the scientists to come
forward—as they are repeatedly urged to do in usually poorly attended
training sessions—and discuss the commercial potential of their scientific work. Responses among the researchers vary widely, from diligent
cooperation in pursuit of patents, as intended by the Bayh-Dole Act,
to indifference, to hostility. At the University of Wisconsin–Madison,
which has one of the smoothest-running and oldest tech-transfer operations in the nation, managing director Carl Gulbrandsen told me that
even today "you see some faculty members who feel this is unseemly."

Asked if these faculty members reject efforts to interest them in tech transfer and patenting, he replied:

> Yeah, periodically, we'll have a faculty member that will say, "We're not going to do anything to commercialize this. In fact, I'm going to publish it [thus jeopardizing patenting]." Well, you talk to people. And this is the exception, but you have a few and for some reason it offends them. And that's their belief structure. You can't do anything about that.[9]

To cope with this challenge, the Bayh-Dole Act provided financial incentives to stimulate academe's innovative spirits. Under the law, inventive professors, along with their laboratories and departments, are entitled to a substantial share of the licensing revenues produced by patents, usually one-third each, with the balance, less expenses, going to the university treasury. Under separate legislation, researchers in government laboratories are eligible for up to $150,000 a year for ten years for income derived from their discoveries. Scientists are thus offered financial encouragement to think of the marketplace as they pursue knowledge and simultaneously provide mentoring and on-the-job training for the next generation of scientists.

Some universities do not rely wholly on the lure of money or the requirements of the law to connect industry to their research. At MIT and several other universities with long experience in commercializing research, matchmaking between campus and corporation is carefully organized and systematically pursued. Karl Koster, MIT director of corporate relations, told me that some 170 companies—each paying $50,000 to $60,000 a year, or more in some cases—are enrolled in MIT's Industrial Liaison Program. Koster explained that in pursuit of corporate deals, the program puts on conferences where faculty members and corporate representatives can discuss their interests. "I think for the faculty, more than anything, it's a chance to hear what industry is interested in and to explain what their particular research programs are. In some cases, it does result in research agreements." He added: "We identify companies that we think would benefit from a relationship with MIT, and we go out and we visit with member companies and we talk to the executives and see if they'd be interested in coming to campus and meeting with faculty." The program, Koster emphasized, is "proactive"; he explained that "we have officers who work with a set of companies, and so they keep an eye out for what kind of research is relevant to the company. Maybe there's a new faculty member and they're working in some aspect of material design or something, and one of the MIT officers will say, 'Well, that will be interesting to my company.' So, they contact the company and say, 'Here's some information on this faculty member and would you like us to set up a meeting with him?' " [10]

But at many universities, including some with extremely large amounts of federally supported research, Bayh-Dole has had minimal effects. The venerable Johns Hopkins University, with over $1 billion annually in federal grants, is a laggard in the academic tech-transfer boom, collecting in each of recent years just a few million dollars in commercialization income. In 1999, in proud defense of his university's commercial backwardness, Hopkins president William Brody asserted that "our scientists are by nature explorers. . . . Asking them to become managers, marketers, and accountants is unrealistic and ultimately inimical to the research enterprise. Time spent in the boardroom is time away from the laboratory, making them less productive and less likely to achieve the things most suited to their abilities." In what elsewhere would be considered a confession of dereliction of duty, Brody boasted: "When Hopkins scientists discovered restriction enzymes, one of the bases of the biotechnology industry, we put the discovery in the public domain—losing millions and millions in potential royalties. Foolish?" he asked. "Perhaps. But I know we didn't slow science down or diminish the leading role [that] American industry plays in this field." [11]

But even with these noble sentiments at the top, the great Hopkins was not immune to the wiles of the marketeers and the temptations of commerce. In 2006 the *Wall Street Journal* revealed that Hopkins had entered into a deal allowing its prestigious name to be used on a line of skin-care products sold by a company in which Hopkins was to receive an equity stake and a board position. Promotional material said the products were tested "in consultation with Johns Hopkins Medicine." Under the glare of publicity, President Brody and the CEO of Johns Hopkins Medicine, Edward D. Miller, disavowed the deal, in lofty terms. Though Hopkins had agreed only to review the scientific validity of tests for the skin-care products, the university's involvement was easily misinterpreted as an endorsement, they acknowledged. "That perception has led some to wonder whether we have allowed financial considerations to overcome long-standing policy separating our work from even the appearance of commercial influence. We can assure you," the president and dean declared, "that is absolutely not the case." To which they added, "At Johns Hopkins, truth, independence and integrity are fundamental to our culture and our academic mission." [12]

Hopkins' retreat from the cosmetics trade was accompanied by the announcement that the university's business dealings would be reviewed by a former director of the U.S. Office of Government Ethics, Stephen D. Potts, chairman of the nonprofit Ethics Resource Center, in Washington, D.C. Score a victory for embarrassment and shame.

The Bayh-Dole Act is widely credited with promoting a flood of innovation from university laboratories to the U.S. Patent and Trademark Office and then on to industry. It appears that way on the face of it, though a small group of economists and lawyers skeptically contends that Bayh-Dole hitched a ride on currents that were already briskly flowing from universities to industry.

Doing Well without Bayh-Dole

With or without exclusivity, patents springing from university research were doing well prior to the passage of the Bayh-Dole Act. The popular sports drink Gatorade, perhaps the best-known commercial product to come out of academic research, was developed for the sweating football team at the University of Florida in 1965—fifteen years before Bayh-Dole—patented, licensed to a food manufacturer, and has produced over $80 million in royalties for the university. The enormously successful Cohen-Boyer technique—crucial for the biotechnology industry—was also patented prior to the passage of Bayh-Dole and was widely licensed on a nonexclusive basis. The offered explanation for Cohen-Boyer's success is that it embodies research techniques for which the market is dispersed among numerous academic and industrial laboratories and manufacturers. But patents, with or without exclusive licensing, are not invariably the legal underpinning for successful manufactured goods. Some companies—particularly in fast-moving information technologies, with their short product cycles—rely on speed to market rather than patents to gain advantage over their competitors. Or, as is frequently the case with computers and other electronic goods that incorporate many patented technologies, competing manufacturers will cross-license their patent holdings as a necessity for producing the goods and quickly getting them to market before a new wave of products emerges. In contrast, after approval by the FDA, pharmaceutical drugs depend on patent protection for a long market life, until patent expiration allows in low-cost generics.

For fending off competition, trade secrets can be as effective as patents, maybe more so, since a patent, in return for the legal protection that it provides, requires public disclosure of the means and methods underlying an invention. That information can enable a competitor to invent a legal way around the patent. The original formula for Coca-Cola is famously guarded as a trade secret, said to be locked in a vault of the Sun Trust Bank in Atlanta. (Whether it really remains secret is doubtful, given the power of modern chemical analysis and the proliferation of taste-alike colas, but the claim is part of the Coke mystique.) Trade secrets stay with the company and are not displayed to the world at large, as is the case with patents, which must reveal the "how" of the invention. But trade secrets come with a risk: Competitors are free to analyze and reverse engineer and go to market with products based on trade secrets. Moreover, trade secrets, though common in industry, are rare to nonexistent in academe, where secrecy in science is anathema, or is supposed to be. However, whether embodied in patents or concealed as trade secrets, basic scientific knowledge is not easily bottled up. Both new scientific understanding and research techniques with commercial value often take wing, legitimately transmitted to industry by fresh hires out of graduate school or following completion of postdoctoral fellowships. Valuable knowledge can transfer over lunch.

Though patents, at least to laymen, convey an impression of valuable property, most patents do not blossom into profitable products. Even in universities with notable success in licensing patents, technology-transfer managers acknowledge that a few among hundreds in their portfolios bring in most of the money. Before its stem cell patents achieved commercial importance, Wisconsin's Carl Gulbrandsen noted that patents related to vitamin D provided 70 percent of the licensing revenue collected by the university's tech-transfer organization, the Wisconsin Alumni Research Foundation. Jokingly, he told me that he has urged WARF's staff to seek other lines of patent income, because "if somebody finds that vitamin D causes cancer tomorrow, we're in a lot of trouble." [13] Given the circumscribed role of patents in the capture of economic value from scientific knowledge, what accounts for the veneration of the Bayh-Dole Act as an elixir of the American economy? The answer is a scattering of impressive financial triumphs, a determined cheering section, and a gullible press.

The Tech-Transfer Profession

When federal law decrees something to be done, careers are born. Founded in 1974, the Association of University Technology Managers (AUTM) rapidly expanded after the passage of Bayh-Dole, in 1980. AUTM comprises university officials charged with selling science produced in their schools, along with various businesspeople and professionals interested in promoting the process and sharing the gains. In AUTM's version of the story, the Bayh-Dole Act is an unalloyed success, undeserving of the reservations and criticisms by theory-blinkered economists and antediluvian purists nostalgically doting on a long-ago gentlemanly era of science.

With universities required by the Bayh-Dole Act to assess the commercial potential in their government-financed research and pursue business deals, university technology-transfer offices sprouted where none had been before, and those already in existence expanded their activities and scope. Individual membership in AUTM has risen from 100 in 1980 to over 3,500 today. In telling the public about tech transfer, AUTM and its boosters validate a tactical rule for success in Washington, namely: Some numbers beat no numbers every time. The champions of Bayh-Dole muster columns of impressive statistics. Between the passage of the Bayh-Dole Act in 1980 and 2005, the number of patents annually issued to universities rose from about 250 to 3,278. During those years, university inventions spawned the creation of 5,171 new companies, of which perhaps half were still in business in 2005—a healthy survival rate for customarily fragile start-ups. Royalties on sale of products based on university patents totaled $1.1 billion. Thousands of jobs were created.[14] Presented annually, the AUTM data were favorably echoed in the press nationwide. Little attention was given to concerns about the downside of academic-commercial links, such as corporate secrecy invading the halls of science, conflicts of interest, monopolizing of important diagnostic tests and therapies, and diversion of scientists and students to trite moneymaking chores.

Income from selling science is a quantifiable indicator of performance in technology transfer, so it is no wonder that in AUTM's public announcements, income received top billing when tech transfer first began to produce substantial financial returns. But with universities increasingly sniped at for veering toward commercialism, rhetoric concerning the goals of tech transfer has recently shifted in the direction of public service, rather than moneymaking, as the prime motivation. Covering the years 1991–95, revenue as the most important outcome of technology transfer emerged tops in a survey of technology-transfer offices and management at sixty-two research universities. Faculty members ranked revenue from tech transfer about on a par with money provided for sponsored research. However, revenue as the most important outcome of tech transfer was unsurpassed by the other

choices—inventions commercialized, licenses executed, sponsored research and patents.[15]

Of late, however, the long-standing emphasis on moneymaking has
been discreetly superseded by claims concerning the public benefits of
tech transfer, as manifested in job creation, local economic growth, and
socially useful products, particularly for medical purposes. The switch
responds to concerns that if universities are seen as behaving like commercial enterprises, the public and politicians will come to regard them
as such, possibly jeopardizing their tax-exempt status and charitable
receipts. Several universities have taken to withholding revenue data
from their annual reports to AUTM, among them Yale and Columbia,
both high in tech-transfer income, published or not. "I don't believe
the measure of our success should be the royalty dollars brought in,"
stated Jonathan Soderstrom, director of Yale's Office of Cooperative
Research. "So this is my little world of protest," he said in explaining the blackout of Yale's income from patents—when last reported,
$40 million, mostly from the HIV/AIDS drug ZERIT. The emphasis
on royalty income, he said, overlooks the positive economic effects that
research at Yale has had on the growth of company start-ups, venture
capital, and real estate values in New Haven.[16] Those benefits of tech
transfer are indeed worthy, but the financial returns are also of interest. So much for "transparency," ubiquitous in the rhetoric of academic
management but now selectively removed from the public record.

Despite the impressive statistics, most universities barely, if at
all, cover their staff and legal expenses in identifying commercially
promising discoveries and patenting and licensing them to an industrial firm or a campus-spawned start-up that can produce something
to sell. However, for the laggards in this difficult business, inspira-
tion is provided by a handful of universities that receive large returns
from their patent portfolios, in accompaniment to their public service.
Occasionally there's a spectacular, one-shot payoff that, like a lottery
super-jackpot, reinvigorates the hopes and dreams of all players. The
biggest ever was announced in 2005: in lieu of prior payments, a onetime royalty of $525 million to Emory University for a widely used
AIDS drug, Emtriva, discovered by Emory researchers and licensed
by the university to two pharmaceutical firms. Also in 2005, Stanford
University collected $336 million from the sale of the Google stock it
received in return for licensing Internet search technology to the company created by two of its former graduate students. Impressive gains
from sales of science are reported by other universities. New York Uni-

versity received $133 million in net licensing income; Wake Forest University, $50 million; University of Wisconsin, $49 million; and University of Minnesota, $47 million. Twenty-five universities each received more than $10 million in licensing income.[17]

Large as they are, these sums are relatively small in comparison to the research budgets of most major league universities, according to AUTM. Wisconsin's research spending totaled nearly $800 million in 2005; Minnesota's, $548 million. Many universities spend enormous sums on research but earn little in return from patent licensing. Johns Hopkins reported over $1 billion in research expenditures but merely $12 million in net licensing income.[18]

Harvard leads in rhetorical qualms about commercialized science. Though it has engaged in technology transfer, it has until recently refrained from the determined pursuit of patents and licensing characteristic of other major universities. Paradoxically, however, Harvard gave birth to one of the biggest, and most controversial, academic-corporate science deals of all time, its exclusive licensing to DuPont of the patent for the genetically modified Onco Mouse, also known as the "Harvard Mouse" and the transgenic mouse, patented by Harvard in 1988. Specifically designed to increase its susceptibility to cancer, the ill-starred creature quickly became an indispensable "research tool" for advancing basic understanding of cancer and for testing pharmaceutical drugs and innumerable other products. DuPont helped finance development of the mouse with a $6 million donation to Harvard. Two other patents followed, and DuPont drove hard bargains in selling the mouse, insisting, among other demands, on "reach through" rights for a share of any profits resulting from its customers' use of the mouse. Under pressure from academics and the NIH, which had also financed some of the research, DuPont agreed to waive charges for university researchers and other nonprofit users. Though sales figures have not been disclosed, the *Scientist* safely referred to the Onco Mouse patents in 2004 as "some of the most valuable pieces of intellectual property ever created"—valuable for DuPont, that is.[19] Harvard's yield has been negligible, if any at all, as can be seen in its consistently low figures for licensing income. Though a major recipient of federal research money—$348 million in 2004—Harvard's licensing receipts amounted to only $16.6 million that year. Corporate sponsorship of research at Harvard totaled only $5.8 million, compared to $72 million at its proudly entrepreneurial neighbor, MIT. Lita Nelsen, head of technology licensing at MIT, explained that "Harvard, if you go way back, was quite ambivalent about whether they should be doing this kind of stuff—soiling the ivory tower with the grubby fingerprints of industry." In 2005 Harvard shed its reticence about corporate dealings, appointing a new chief of technology development, Isaac Kohlberg, a veteran of academic-commercial enterprise; it also created a $10 million fund to promote technology transfer. Kohlberg announced that Harvard's technology-licensing employees would henceforth be known as "directors of business development." Harvard provost Steven Hyman was quoted in the *Boston Globe* as saying, "Our mission demands that beyond writing scientific papers, that technologies be commercialized so they can make a difference for people." [20] Harvard's tardiness in arriving at this ubiquitous belief was not explained.

The patenting business that routinely thrives on America's campuses receives overwhelmingly laudatory attention in the popular press, with little or no skeptical scrutiny or inquiry about the actual profits and losses, in dollars and academic and scientific values. A rare exception appeared in *Fortune* in 2005, in an article, "The Law of Unintended Consequences," that attributed the slowdown in pharmaceutical drug innovation to erosion of the "scientific commons" and an epidemic of patent litigation—both inspired by Bayh-Dole's mandate for universities to seek patents and profits from their government-financed science, according to the author. [21] The explanation for the predominantly pain-free journalistic treatment is that news about academic tech transfer comes almost entirely from members of AUTM and their cheerleaders in university public relations departments, in collaboration with university boosters. And it's always welcome news, about the good work of dedicated scientists being conveyed to industry and eventually to the public. And making it even better, the patents bring financial benefits to the university and its scientists and provide money for more good work by them.

"The Most Inspired Legislation"

Miraculous economic effects have been ascribed to the Bayh-Dole Act. In 2002 the *Economist* gushingly described it as "possibly the most inspired piece of legislation to be enacted in America over the past half-century," adding, "More than anything, this single policy measure helped to reverse America's precipitous slide into industrial irrelevance." [22] The unalloyed encomium warrants examination. Over several decades to the present, universities have received only a sliver of all U.S. "utility" patents, the most frequently issued type of patent. In 2004, when some 3,200 patents were issued to members of AUTM, nearly 85,000 utility patents were issued in the United States, mostly to corporations, and another 80,000 were issued for foreign inventions.[23] Over many years, the huge disparity between academic and corporate patent totals has prevailed. The intellectual quality and economic impact of the academic patents might possibly be superior, but evidence for that possibility is lacking. And even with the substantial increase in patenting, the proportion of academic scientists actively involved in science for sale remains relatively small and mainly confined to a few fields, large among them biotechnology—a sinkhole for a great deal of investment money, rather than a fount of prosperity. In 2006 the president of Genentech, one of the very few profitable firms in the industry, described biotech as "one of the biggest money-losing industries in the history of mankind," with losses of over $100 billion since 1976.[24]

A "Gold-Digger Mentality" in Academe?

Critical evaluation sometimes penetrates the PR curtains. In 2005 the prior enthusiasm was nowhere evident in the *Economist*'s twenty-fifth anniversary assessment of Bayh-Dole. While recalling its earlier celebration of the law and its impact, the magazine now reported that

> the critics have grown louder over the years. Many scientists, economists and lawyers believe the act distorts the mission of universities, diverting them from the pursuit of basic knowledge, which is freely distributed, to a focused search for results that have practical and industrial purposes. . . . What is not in dispute is that it makes American academic institutions behave more like businesses than neutral arbiters of the truth. . . . Moreover, there is ample evidence that scientific research is being delayed, deterred or abandoned due to the presence of patents and proprietary technologies. . . . Even industry is beginning to complain about a gold-digger mentality among academic administrators.[25]

Such commentary is rare. Increasingly, however, AUTM exhibits a surprising degree of insecurity about public perceptions, though it is doubtful that many in the public are aware of the Bayh-Dole Act or its effects on academic science. The critics, however, are no match for

AUTM, which has developed into a confident and mature nationwide organization that trumpets the value of academic-industrial linkages. Having outgrown its crass-sounding celebration of dollars as the measure of success, AUTM currently defines its raison d'être as advancement of the common good via the delivery of new technologies, jobs, and cures for the American people. Moneymaking is deemed incidental though welcome. "Many people are often confused about why we are interested in technology commercialization, in nurturing start-up companies, and in facilitating more patents and license agreements," Mary Sue Coleman, president of the University of Michigan, told AUTM's 2005 annual meeting, which celebrated the twenty-fifth anniversary of the passage of the Bayh-Dole Act. "It's not about the promise of future revenues that might be generated from this activity. You heard me correctly," she insisted. "It is not about the money. . . . Technology transfer must serve our core mission: sharing ideas and innovations in the service of society's well-being." [26]

The emphasis on good works rather than moneymaking was particularly evident on the silver anniversary. In observance of the occasion, AUTM inaugurated "The Better World Project," which aimed "to promote public understanding of how academic research and technology have changed our way of life and made the world a better place." The following year brought the publication of *The Better World Report: Technology Transfer Stories: 25 Innovations That Changed the World.* Listed among the inventions by AUTM's university-based researchers were Google, the V-chip for blocking unwanted TV programs, the Honeycrisp apple, and the PSA test for prostate cancer. Included, too, was a message from the coauthor of the act, former senator Birch Bayh, now a lobbyist in Washington, who assailed the critics of his legislative creation:

> The modern-day detractors of Bayh-Dole, who suggest that this legislation creates an incentive for researchers to get rich, which is more important to them than honest research, have no understanding of what motivates those who devote their lives to science. There may be a few greedy researchers, however, the odds are stacked heavily against a scientist living on easy street. The great motivating factor in their lives is expanding the field of human knowledge, coupled with a passion that their research finds a practical application. . . . It is unfortunate that today's critics of Bayh-Dole spread the belief that all

problems which may exist in those instances where they allege wrongdoing can be remedied by making the product of university research available to the public generally. This principle sounds good in a vacuum, but it has failed dismally in our free enterprise system. This was the status before Bayh-Dole, when more than $30 billion of taxpayer money had been spent on research, only to produce thousands of patents gathering dust in the Patent and Trademark Office.[27]

Adulation from Abroad

AUTM's soaring statistics of patents and jobs have generated admiring worldwide attention. In part, this may be attributed to the penchant of scientists and their bureaucratic patrons in the granting agencies to look abroad and warn that their country is falling behind in science and its applications. American scientists and their patrons are old hands at this alarmist game. But in tech transfer from academe to industry, the United States is notably advanced. From abroad, government, industry, and academic officials flock to the United States to imbibe the secrets of tech transfer from leading academic practitioners. "They come by and try to find out how Stanford does it," Katherine Ku, director of Stanford's Office of Technology Licensing, told me. From all over the world, she said, the traffic is so heavy and continuous that "we've asked people to come in at Friday at ten o'clock. And it will be foreign delegations from literally all over. Norway, Sweden, Finland, Belgium, England, Germany, France."[28] While the rhetoric of Bayh-Dole's cadres has evolved past moneymaking as a prime motivation, the aspiring foreign emulators of America's success frankly acknowledge the attraction of money. A 2003 study of "Business-University Collaboration," conducted for the UK's Treasury Department, noted an increased interest in tech transfer among Britain's universities and observed:

> This trend has been driven in good measure by money. Universities have been forced by economic circumstances to hunt around for new sources of cash and equipment, putting a new emphasis on business partnerships. Third stream funding [support for tech transfer, in addition to support for teaching and research], although relatively modest in size, has provided an incentive to build relationships with business. In addition, the

development of new science-based industries—especially in biosciences and information technology—has created fresh opportunities for researchers to work with business. A new role model, the entrepreneurial academic, has appeared on many campuses and some of them have become quite rich as a result of their efforts in consultancy, or by creating and subsequently selling spinout companies.[29]

Britain's University of Cambridge, a scientific powerhouse in its own right, enthusiastically partners with MIT in a transatlantic program that exchanges over fifty students each for a year of study focused on academic entrepreneurship. Cambridge is a leader in the UK's efforts to promote tech transfer, and learning from each other's experiences is the avowed goal of the two institutions, but it's well understood that MIT is the master in this field.* The champions of Bayh-Dole take heart from the admiring foreign interest. Summarizing and deploring allegations that Bayh-Dole, as she phrased it, "has fundamentally changed the nature of U.S. academia for the worse," Ann Hammersla, MIT senior counsel for intellectual property, retorted in her capacity as AUTM president in 2004–5:

> Ironically, this U.S. introspection is underway even as the magnitude of the United States' success has garnered respect around the world, and many countries are changing their university and legal systems and making substantial financial investments to emulate the environment that Bayh-Dole has created. . . . These [technology] transactions provide funds that institutions reinvest in research and education. But more important is the impact that they've had on society. These products improved people's lives, and spurred new jobs at the companies that developed and now sell them. Technology transfer is about passion—in the invention and its development and implementation.[30]

Technology transfer originating in universities engages a far-flung professional and business constituency. Of AUTM's membership,

---

*Entrepreneurial spirits at Cambridge were strong but loosely harnessed to the institution until 2006, when—over strong protests—new rules were adopted giving the university first option on inventions in its laboratories. Prior to the change, staff scientists were free to patent and license their discoveries and exclusively reap the financial benefits.

45 percent are employed by universities; 13 percent are lawyers, mostly patent attorneys in private practice trolling for business as outside counsel to universities (since patenting, licensing, and associated litigation are specialized and intermittent, university tech-transfer offices typically employ outside patent law firms); 11 percent are from industry, seeking marketable research in university laboratories; and 6 percent identify themselves as consultants, which covers a broad range of activities that bring together academe, finance, and industry. The remainder are scattered among foundations, hospitals, venture capital firms, and other organizations.

Like many professional associations, AUTM exploits a valuable possession: access to its members, available at a price. At a 2004 regional meeting in Charleston, South Carolina, for $14,000 companies were invited to purchase the opportunity to sponsor a "continental breakfast, luncheon and refreshment break . . . to increase the cost-effectiveness and impact of your marketing plan." The sponsors on that occasion included the AstraZeneca pharmaceutical firm and a law firm specializing in patent work.

In the organizational structure of academe, specialists in technology transfer hold an uncertain status. Many possess advanced degrees and previously moved between industrial jobs and university teaching positions. But in their university tech-transfer roles, few hold faculty appointments; the tech-transfer office is not part of traditional academic administration, leaving them in undefined territory. Moreover, to a varying extent, depending on the school, the tech-transfer staff's unabashed concern with commercial matters does not harmonize with the ethos of academic research as an enterprise aloof from profit-seeking. Insecurity about status was evident in corridor chatter and a formal presentation at the 2004 Charleston meeting, where an AUTM member reported that at his university the tech-transfer office sought adjunct teaching appointments for its staff members as a means of raising their status on campus. "It makes people feel good," Mark Crowell, director of technology transfer at the University of North Carolina, explained, adding that adjunct appointments "give stature and credibility to people sometimes perceived by faculty as lesser members of the university." [31]

AUTM leaders proudly assert that academic tech transfer is evolving as a profession, that it is taught in university courses and is so specialized and demanding of unusual combinations of skills that headhunters are employed to recruit staff. Talent alone is not enough. Of great importance, too, tech-transfer specialists told me, is the attitude at the top of university administration. At Georgia Tech, where technology transfer is a campus-wide holy cause, George G. Harker III, director of the Office of Technology Licensing, explained that presidential support, a clear commitment to commercialization, and, especially, a willingness to take business risks are crucial for success in tech transfer. Recalling visits to other universities to discuss tech transfer, Harker said, "I have asked the president how risk averse are you? And I know they're going to say, 'No problem.' I look at the way they say it, because if they're really risk averse, the tech-transfer office is going to be stifled. If they're going to lie awake at night at every deal you make, the deals are going to be very limited." Harker stressed that the "'alignment' of the university is very critical for success." By "alignment," he explained, he meant top-to-bottom support for tech transfer.[32]

The daunting odds for success in tech transfer were emphasized to me by John A. Fraser, director of technology transfer at Florida State University, home of one of the rare "blockbusters" in technology transfer, the cancer drug Taxol. But overall, he said, tech transfer confronts dismaying prospects for success:

> Crummy business. Crummy business. You put this up in front of an MBA class, and they'll laugh you out. Because the numbers are against you. At Stanford they did a study over their thirty years of existence [in tech-transfer activities] and showed that 50 percent of the deals done at Stanford brought in ten thousand bucks or less. Stanford! If you go to the president of the university, and the president says, "I want to support an activity in this institution to raise money for my institution," I would not be the first one there on line. Of the 125 [blockbuster patents among all universities], this university has been most fortunate to have had a major hit. But there are studies coming out that show more tech-transfer offices than not in the nation, until you get up to eight or ten years of life, don't stand a chance of break-even, on average. If you're a small university, maybe after twenty years, still maybe not. So why do this? There are several reasons. It's mandated by Bayh-Dole. It's an expression of creativity, which faculty are getting very, very interested in for a variety of reasons. It touches on areas that the university and its faculty are interested in: working with the private sector. And it also turns out to be an area where it's fraught with some problems.[33]

The NIH and Tech Transfer

Success in the tech-transfer markets depends not only on the subject matter and quality of an institution's research activities, but also on its ideological and psychological comfort with the pursuit of commercialization. Nowhere is this more apparent than at the organizational and financial center of the federal government's biomedical research enterprise, the National Institutes of Health, budgeted for a colossal $28 billion in 2005. The NIH is renowned for the billions that it annually provides for research by nongovernment scientists, in universities, medical centers, and freestanding research institutes. But the NIH also runs its own research projects, conducted by government employees, in what is known as the NIH Intramural Program. Mainly situated on a 300-acre campus, crowded with laboratories, clinical research facilities, and administrative buildings, in Bethesda, Maryland, on the outskirts of Washington, D.C., this is an immense enterprise. About 10 percent—$2.8 billion in 2006—of the NIH's budget is spent on intramural research, making the NIH by far the biggest and richest biomedical research performer in the world. With some six thousand intramural scientists at work on two thousand research projects at any one time, the NIH ceaselessly produces a flood of scientific papers that achieve a high rate of publication in leading journals. NIH scientists, too, are bound by statutory imperatives to go commercial. Yet only a trickle of technology runs from the NIH to the pharmaceutical and biotech industries. In 2005, from thousands of published research papers, NIH scientists made only 388 invention disclosures, from which 186 patent applications were filed. In that year, 62 patents were issued to the NIH, and royalties from licensed patents amounted to $98 million—all strikingly low figures, given the large number of projects, published papers, and the dollar volume of research conducted by NIH scientists.[34]

The explanation for this paltry performance is complex and is best summed up as the product of political skittishness in the NIH culture compounded by old-time scientific aloofness from the market. From Capitol Hill, vestiges of pre-Bayh-Dole populist suspicions are directed at the NIH, with accusations that the NIH—ineptly or intentionally—engages in giveaways of taxpayer-financed research to pharmaceutical firms, which use the knowledge to produce drugs that are sold to the public at exorbitant prices. The discovery, development, and marketing of AZT, the first effective drug against the AIDS virus, has been cited

as an egregious example of the NIH's disregard for the public interest. AZT, short for azidothymidine, was first synthesized at Wayne
State University in 1964 by Jerome P. Horwitz, a chemistry professor
who was seeking anti-cancer drugs. When the compound proved disappointing for that purpose, Horwitz set it aside and went on to other
research—a common course of events in the uncertainties of scientific
investigation. In the mid-1980s, however, as the spread of HIV/AIDS
inspired urgent searches for treatment, scientists at the National Cancer Institute, a part of the NIH, tested many compounds initially developed for other purposes and found that AZT slowed the growth
of the deadly virus. Burroughs Wellcome, since merged into Glaxo-
SmithKline, worked further on the drug, patented it, and turned it
into a pharmaceutical blockbuster, priced in its early days at $8,000 to
$10,000 for a year's treatment. The price has since come down, but
AZT remains a big earner, registering worldwide sales of $1.7 billion
in 2004. Horwitz, who did not patent the drug in those long-ago days
of commercial innocence in academe, received nothing, beyond sym-
pathetic press reports of his invaluable but unrewarded discovery. He
later accused the NIH of a giveaway to the pharmaceutical firm, telling
the *Chronicle of Higher Education*, "There was no reason to award
that patent to Burroughs Wellcome in the first place."[35] Burroughs
Wellcome and GlaxoSmithKline prevailed in lawsuits brought by ge-
neric drug manufacturers challenging the patent for the lucrative drug,
but the litigation may not be over.

A Handout for Industry?

The NIH's role in the circuitous route to discovery was easily portrayed
as heartless complicity in price gouging for a life-or-death drug developed at public expense. "AIDS Drugs: Is the Government Research
Program a Helping Hand for Patients, or a Handout for the Pharmaceutical Industry?" was the title of an article in 1989 in *Health Letter,* published by the Nader-related Public Citizen Health Research Group. In
the same article, an answer, and the dilemma inherent in the question,
was provided by Anthony Fauci, head of AIDS research at the NIH:
"When taxpayer money goes into the development of a drug, then it
should not be sold for an outlandish price. On the other hand, if you try
to have government regulation or encroachment on the rights of various companies, you may discover they are not interested."[36] The AIDS
population, well connected to the arts and entertainment, and deft at
public relations, angrily denounced the pricing as unjustified for a drug

initially developed at public expense. The characterization was disputed by Burroughs Wellcome, which said it bore the heavy costs of development after the NIH had discontinued research on the drug. But in the public mind, veracity and the pharmaceutical industry are not a twosome. Within the NIH, the preexisting doubts about the propriety of commercial links were heightened by the AZT episode and the backlash it suffered from its dealings with Burroughs Wellcome. The agreed-upon facts of the convoluted episode can be arranged to support conflicting conclusions: that absentminded bumbling by the NIH accrued to the benefit of a sharp-eyed, profit-driven pharmaceutical firm, or that the NIH, embedded in a capitalist system and operating within the conservative Reagan administration, harnessed the profit motive to speed a lifesaving drug to thousands of otherwise doomed patients. Either way, the experience did not enhance the NIH's appetite for commerce.

The NIH has also been battered by accusations that it virtually gave away to Bristol-Myers Squibb (BMS) the rights to what eventually was marketed as Taxol, the best-selling anti-cancer drug of all time, with over $9 billion in sales worldwide from 1993 to 2002. (We'll examine the Taxol saga in detail in a lengthy conversation with its discoverer, Robert Holton, in chapter 8.) For the ideological descendants of Bayh-Dole's original opponents, the Taxol story was further evidence of officially sanctioned hijacking of government-financed science. In a regular feature titled "Outrage of the Month," the *Health Letter,* published by the Public Citizen Health Research Group, headlined a report: "Taxol: How the NIH Gave Away the Store." The article asked:

> What does the NIH (and the public) have to show for all these years of government creativity and investment? A paltry $35 million, according to the GAO. Even this minimal cost to BMS has been recouped many times over from the Federal government itself: through its Medicare program, the U.S. spent $687 million on Taxol between 1994 and 1999.[37]

Steven M. Ferguson, longtime head of technology transfer at the NIH, told me that "this office probably got a real impetus because of the problems NIH had with Burroughs Wellcome and AZT, which at the time that was happening, there was not really any substantive program that could have reviewed that." Under the glare of indignant legislators and public interest organizations, the NIH now pays very close attention to industrial deals and licensing terms. The political atmosphere surrounding such dealings became further sensitized in 2003 when the *Los Angeles Times* disclosed the existence of hidden, lucrative private consulting deals between scores of senior NIH administrators and pharmaceutical and biotechnology firms. But also in the picture, Ferguson told me, is a lingering, though diminishing, reluctance by many NIH researchers to engage in any aspect of commercialization. Attention to technology transfer, he said, with its requirements of

> disclosure, is seen by some NIH researchers as another bit of bureaucracy, and it gets in the way of science; or researchers say, "Please stop bothering me, I'm here to treat patients, or I'm here to do my science." What's coming up is a sort of gradual appreciation that technology transfer is actually an integral part of the process, particularly when you think a little further downstream of NIH's role as a health agency. I think you're realizing that some of the things we need to do as a health agency are related to products coming on the market from re-search, not necessarily [only] scientific publications, because the actual taxpayer doesn't necessarily relate to a scientific ar-ticle, but would relate to a new medicine or a new diagnostic test that their relatives could use or they would have access to. That, and also the idea that some of the scientists are actually realizing that there is importance or scientific satisfaction in seeing their work commercialized.

But even with those recognitions, Ferguson acknowledged, changes in attitude toward commercialization come slowly. "NIH hasn't really wanted to market itself," he pointed out, noting:

> What we don't really have is more the entrepreneurial type of culture, where the scientists can have start-ups and that sort of thing. We've avoided some of those problems, but we've had our own problems and other issues. But partly that is because we've been held to very high standards. And I guess in the end we can't complain about that. What do people really want from us? Do they really want more patents, more products, more companies down the street that are collaborating with us? Spin-offs, whatever? That kind of thing is being sorted out. We kind of get mixed messages. Part of it right now is that the NIH is above all that. It's still the last ivory tower—some days. But other days of the week, we're right mixed in it, and criti-cized for collaboration. So, it depends what day of the week.[38]

From the perspective of the biotechnology industry, initially a creation and still an intellectual dependent of academic science, the NIH's ambivalence toward tech transfer is deplorable. Chuck Ludlam, an attorney, served from 1993 to 2000 as vice president for government relations of the Biotechnology Industry Organization (BIO), the Washington-based lobby for the industry, and later as counsel to Senator Joe Lieberman (D-Connecticut). Ludlam speaks venomously of the NIH's attitude to tech transfer and especially of efforts in the 1990s by congressional liberals, led by Senator Ron Wyden (D-Oregon), smarting from the AZT and Taxol experiences, to invoke a "reasonable pricing clause" for government purchase of drugs derived from NIH research. And he's frank about the profit-seeking fervor in the industry. The companies in BIO were enraged by the NIH, Ludlam recalled:

> Ninety-five percent of my members said they would never even talk to NIH as long as the reasonable price policy was in effect. Because they do not believe in reasonable prices. They believe in 30 percent operating margins, or whatever else they can possibly get, because they are in a very risky business, where they have very high expenses, very high costs of capital, and greedy investors. So the idea that the government would be reviewing their prices in any way was the end of the conversation. At one point in this process, I did a survey of all the tech-transfer programs in the government—Department of Energy and all those other agencies. And by any possible standard, NIH had the *worst* one—the most bureaucratic, the most delay, the most risk averse, the least practical, the most ideological. The whole gist of HHS [Health and Human Services, departmental parent of NIH] is price controls, anti-industry, anti-patent. It pervades the entire department, including NIH. NIH has the worst record of licensing of any agency in the government. They have a pathetic return on their investment in terms of royalties. You can't get a license out of them in less than a year. It's going to have ten extra clauses in it. They're anti-industry; they hate industry. Nobody in their right mind would even say that NIH is even remotely in the ballpark with Stanford and MIT and the better university tech-transfer programs. They're incredibly risk-averse—Wyden-averse.[39]

Appointed director of the NIH in 1993, Harold Varmus inherited the reasonable-pricing controversy and treated it like a live bomb.

"The fact is that we had no way to do reasonable pricing," he told me. "We're not regulatory. So I just thought we ought to drop it. We can't enforce it anyway, so why bluff?" Regarding the NIH's sparse record of successes in commercialization of research, Varmus acknowledged that "NIH has not had a big winner"—apart from the blood test for HIV/AIDS, for which the NIH shares royalties with the French Institut Pasteur. "I agree in one sense," he added, "that given the size of the NIH, you might have expected they would have a blockbuster, but it didn't. And if you look at the average across the country, there are a lot of good institutions out there—we can all rattle off the six or seven: Columbia and MIT and Harvard and Stanford and UCSF and us [Memorial Sloan-Kettering]—who have blockbusters. But the rest don't. What does UCLA, UC San Diego, what do they bring in? Nothing. Pennsylvania too."[40] Varmus's off-the-cuff assessment is not altogether accurate, but, in general, he's correct. Many universities with big R&D budgets reap little from patenting.

The movement of research from academic and government laboratories to industry is an intricate enterprise, affected by differing motives and varying degrees of interest, enthusiasm, negotiating skill, marketplace opportunity, luck, and a generally low success rate.

Bayh-Dole's Skeptics

Doubts about the economic impact of Bayh-Dole and allegations of the legislation's deleterious effects on scientific progress have festered in economic and scientific circles for many years. But even with rising concerns about science losing its soul and perhaps some momentum to commerce, technology transfer draws little public or political attention. The patenting and commercial exploitation of discoveries that are useful for conducting research rather than for producing goods for the general marketplace is increasingly an irritant among scientists. These so-called research tools, like the patented Onco Mouse, are useful mainly for scientists and can be indispensable to the progress of research, but when entangled in patent rights and priced to raise revenue, can delay or block scientific investigations. The danger of patents impeding research was gingerly acknowledged by the President's Council of Advisors on Science and Technology in 2003. Discussions of research tools "need to be monitored" to insure a balance between profits and science progress, the council stated as number ten on a list of recommendations. The number one recommendation: "Existing technology-transfer legislation works and should not be altered."[41]

There is no pressure for changes in Bayh-Dole or even a review by a congressional committee or a specially appointed public body. But if one were to be held, the proceedings would be dominated by AUTM's enthusiastic leaders and their university superiors. "The beauty of Bayh-Dole is that it's been so simple," explained Lita Nelsen, director of MIT's Technology Licensing Office. "And the horror that we all live with is that it's going to be 'fixed.' If it ain't broke, don't fix it," she said, adding that "it will be fixed with a bunch of special interests, and then it will start to look like the tax code." Noting that the act was passed twenty-five years ago, Nelsen said, "There's nobody left in Congress that remembers why it was put in. And it would be subject to all sorts of pork and god knows what else." [42] In 2004 I asked congressional staff members on the relevant committees about the possibility of hearings on the twenty-fifth anniversary of Bayh-Dole. Few had heard of the legislation.

From an economic perspective comes an indictment larded with skepticism about the benefits attributed to the legislation, produced by four academics: David Mowery, University of California, Berkeley; Richard R. Nelson, Columbia University; Bhaven N. Sampat, Georgia Institute of Technology, and Ardvids A. Ziedonis, University of Michigan:

> We believe that much of the current discussion of the economic role of U.S. research universities and the contributions of U.S. universities to the economic boom of the 1990s exaggerate the role of Bayh-Dole. In fact, U.S. universities have been important sources of knowledge and other key inputs for industrial innovation throughout the twentieth century, and much of this economic contribution has relied on channels other than patenting and licensing. . . . The widespread belief held by many policymakers and university administrators in the United States and elsewhere that Bayh-Dole has been an unmitigated success is based on little evidence. First, data on the growth of U.S. universities' patenting and licensing activities alone provide no basis on which to conclude that patenting and licensing are essential for technology transfer, since increased university patenting may cover technologies or inventions that previously were transferred via other channels. Second, increased academic patenting and licensing, as well as growth in other forms of university-industry collaboration, predate Bayh-Dole. Third, the "evidence" on low rates of commercialization before Bayh-Dole is weak. [43]

Noting the award of patents for research techniques, as distinguished from patents for technological developments that can be embodied in marketplace products, the economic assessment warns that "'privatization' of knowledge inputs that formerly were part of the 'scientific commons' through patenting may impede the progress of research. Increased academic patenting also may enhance incentives for faculty or universities to delay publication, restrict sharing of research materials, and/or limit the sharing by faculty of their research results with the scientific community via conference presentations or informal communications."[44]

A stronger expression of similar views comes from the Royal Society of Great Britain, where the scientific tradition of openness and collegiality is in a losing competition with the government-endorsed goal of "wealth creation" derived from academic science. In a policy statement issued in 2003, *Keeping Science Open: The Effects of Intellectual Property Policy on the Conduct of Science,* the organization disapprovingly noted the widening and lowering of patent standards in the United States and a

> growing tendency towards pushing the boundaries of patenting out from inventions into areas of knowledge. . . . Much rhetoric in the US has tended to regard patents as an almost absolute or natural right for inventors. By contrast, in Europe, patents are regarded less as an absolute right than a privilege granted at the discretion of governments in pursuit of economic, social or technological objectives. . . . It is of particular importance to the scientific community that modification to these exclusions from patentability do not lead to a greater risk of scientific knowledge being monopolised. We agree with the view of many scientists that pure knowledge about the physical world should not be patentable under any circumstances. That it should be freely available to all is one of the fundamental principles of the culture of science. Only by having knowledge unencumbered by property rights can the scientific community disseminate information and take science forward.[45]

Under the title "Not Wicked, Perhaps, but Tacky," an editorial by *Science* editor in chief Donald Kennedy gently condemned the downward evolution of ethics and collegial manners in contemporary science. Noting extravagant claims that have suspiciously boosted the share prices of biotech firms, self-serving publicity stunts by scientists, and intransigence in sharing research reagents and other materials, Kennedy stated:

> So what we have here is a growing list of behaviors that, taken together, exemplify the gradual retreat from generosity and straight dealing in a community that is usually known for those qualities. Perhaps the core elements of "tacky" in these examples is that they all eat away at the sense of community, shared understanding, and public trust that are crucial to science.

Kennedy embellished his published remarks with a glossary of definitions of "tacky," among them: "common," "shabby," "seedy," and "marked by a lack of style or good taste."[46]

Even if the Bayh-Dole Act gave academe a vigorous push toward bedding down with industry, there's ample evidence that many professors, with or without the collaboration of their institutions, were headed there anyway. Long prior to passage of the act, some universities were already staffed and on alert for opportunities to sell their science to industry and were doing brisk business. An outstanding example is the University of Wisconsin, which got into the tech-transfer business in 1925 with the creation of the Wisconsin Alumni Research Foundation (WARF) to license university-held patents initially derived from the synthesis of vitamin D, a historic development that led to the elimination of rickets, a disease of malnourishment. Over the next eighty years, WARF received over three thousand disclosures from researchers at the university, obtained some one thousand U.S. patents, and returned $800 million to the university. Stanford, MIT, and many other universities had awakened to the lucre in their laboratories long before Senators Bayh and Dole came to their assistance. And even with the crazy quilt of government patent rules and regulations, the results of government-financed research were getting to market, sometimes protected by patents, but in many other instances through scientific and technical publications, conferences, and person-to-person technology transfer.

When the Bayh-Dole Act went into effect, important discoveries in the life sciences, electronics, and other fields were already spilling out from campus laboratories into existing corporations and professorial start-up companies founded to develop the discoveries. Frisky billions in venture capital financed new crops of companies and along the way created a new academic class of cap-and-gown millionaires. A great deal of academic enterprise was happening anyway for which the

Bayh-Dole Act deserves neither credit nor blame. But Bayh-Dole put a clear federal stamp of approval on the sale of results of government-financed research.

Amid new scientific riches, entrepreneurship, and academe's usual hunger for money, Bayh-Dole was indeed a propulsive force for academic patenting and licensing and for the founding of companies by professors and their universities. Whether it was the matchmaker that brought together academe and corporate America for enterprise that otherwise would have not occurred is doubtful. The AUTM chorus has no uncertainty that Bayh-Dole made the difference. Mowery et al. conclude differently:

> The contributions of U.S. universities to economic growth and innovation during the 1980s and 1990s were important, but no evidence suggests that these contributions were more important than they were during the 1930s or 1950s. Nor does any evidence "prove" that Bayh-Dole substantially increased these contributions or that any such expansion would not have occurred in the absence of the Act. The nature of these contributions and the channels through which they have been realized before and after Bayh-Dole have been complex and have included much more than patenting and licensing.[47]

The complexities may be even greater and less well understood than polemics, scholarship, and journalism have led us to assume, according to an analysis of the origin and growth of solid-state technology at Stanford University. The technological flow was not, as commonly assumed, from campus to corporation, according to Christophe Lécuyer, a research historian at the Beckman Center for the History of Chemistry. Rather, he writes,

> the rise of the solid state electronics programme at Stanford was made possible by massive transfers of technology from industry. Stanford owed much of its competence in solid state circuit and system design to Bell Telephone Laboratories. Its processing expertise came from two companies, Shockley Semiconductor and Fairchild Semiconductor. . . . These flows of ideas, knowledge, and techniques revolutionized the electrical engineering curriculum at Stanford. They also facilitated major research projects on medical instruments, reading aids for the blind, process simulation programs and computer

architectures. . . . Firms such as H-P encouraged the building of the solid state electronics programme at Stanford in the
1950s and 1960s. . . . Industrial transfers enabled Stanford to
systematize and "normalise" knowledge produced in the corporate world. The University also trained students in innovations already developed by industry.[48]

Have we been blind to similar traffic patterns between universities
and the pharmaceutical industry and other life-science organizations
in the private sector? While arguments persist about the importance
and effects of tech transfer from academe to industry, perhaps we have
here an overlooked reality sitting in plain sight: the knowledge flow
runs in two directions.

Prefabricated notions persist in the contention arising from entrepreneurial science, with the corporate branch depicted as a lesser breed,
dominated by profit-oriented goals and timetables, while academic science is the virtuous kind—or would be but for commercial contamination. But the structure of modern-day science has been evolving
in ways that are not always reflected in partisan fusillades. Working
scientists realize this, and it clearly emerges in an insightful study by
two sociologists—neither with great affection for academic entrepreneurship—who conducted interviews at six universities and fourteen
biotechnology firms in 2001–2. In what they describe as "an image of
institutional change that is saturated with contradiction and irony,"
they report that while academe has absorbed characteristics of commercial enterprise, "work in many science-intensive firms is characterized
by a cooperative spirit and freedom from managerial pressures that researchers in today's universities seldom enjoy." Specifically, they noted,
corporate scientists "often find that the conditions and resources needed
to support traditional academic norms—the sophisticated equipment,
the most generous budgets, the greatest distance from entrepreneurial
pressures—are most readily available in corporate laboratories." And,
they concluded, "that in some settings at least, *companies have become more 'academic' than academia itself* [italics in the original]."[49]

While pining continues here and there for a pastoral era in science—of questionable historical reality—the pursuit of money from
knowledge is an everyday reality in the conduct of modern research.
It is not all-pervasive; some scientists remain aloof from commercialization. But tech transfer and other moneymaking opportunities are
now woven into the social system of science, and scientists who once
frowned on business ties now acknowledge a change of mind.

4   Changing Attitudes

We had one of the best biological universities in the world, and why shouldn't we take advantage of that to build exciting young companies? And we had all these great people that we had trained and they were all going to Cambridge and San Francisco. And so this brain drain—this was a good argument to the state. And so now there are lots of jobs, and as a result we're retaining all the people that we've spent time and money helping to train. I actually do care about the economy of the state. The university is supported by the state. We're here and I like it here, and I'd like to do things that are relevant and important to Wisconsin and the country and the world, for that matter. But there's nothing wrong with us wanting to see the basic research here ending up creating businesses and jobs here. I started getting used to the idea of being an entrepreneur. I wasn't one yet. And then a colleague of mine wanted to know whether I'd be interested in starting a company with him. And we decided to do it. This is a company that presently has four scientists and two businesspeople out at the research park. Our whole idea is to engineer a good bacteria to kill a bad bacteria. I'm a scientist who has done a lot of good basic research, but I've never cured anybody's infection. If I can come up with a fundamentally new way of treating infectious bacterial diseases, that's actually pretty important. It's worth testing what we think are really, really exciting ideas. And the only way to do that is to start a company and do it. I have to say, twenty years ago, there was a negative connotation to working with a company. It was sort of you've sold out, you're no longer pure. I don't see that.

Richard Burgess, professor of oncology,
University of Wisconsin[1]

: : :

The amount of activity has greatly increased, and so has the complexity of the deals, but there's nothing new about professors selling their expertise off campus or performing industry-financed research in their university laboratories. They have been doing it at least since the Morrill Act of 1862 and follow-on legislation established the nationwide system of land-grant colleges, with a mandate to promote "agriculture and the mechanical arts." Utility, practicality, and help for the economy were legislated into the culture of many of the great public universities, and many private universities followed that course on their own. Consulting for business and industry long ago became a permissible moonlighting activity in the scientific sectors as well as in other parts of universities, with a prescribed limit of one day a week fairly standard throughout academe. Research projects financed wholly or in part by industry have long been conducted in university science and engineering departments, as have clinical trials in university-affiliated medical centers.

But the volume of such dealings as a portion of overall academic activity remained relatively small until the late 1970s and early 1980s. It was in that period that linkages between academic science and industry rapidly expanded as government money poured into university laboratories, making them founts of industrially valuable knowledge. Money-making excited academic managers, and, as a result, in recent years the tech-transfer movement and other commercialization activities have made a striking transition from a modest existence to a prominent place in the academic world. The phenomenon of professors starting their own companies—the spin-off syndrome—was abetted by revolutionary developments in the life sciences and information technology. Well-established industries existed to absorb the discoveries of inventive professors in other fields. Not so for the pioneers of the gene-manipulating technologies, software, and computers. The traditional pharmaceutical industry, often intellectually unadventurous and financially conservative, viewed the new life sciences as too distant from product development to justify substantial investments. And few existing corporations were sufficiently perceptive to recognize the importance of the electronics developments in garages and lofts around Stanford University and other academic centers. For entrepreneurial professors, the alluring course was to establish a company—often it was the only course for transforming their discoveries into useful products.

Cautiously remaining on the scientific sidelines, the big pharmaceutical firms adopted a vulture strategy of buying biotech start-ups that appeared to be en route to moneymaking products. And, as this pattern became established, many of the major pharmaceutical firms reduced or finally opted out of scientific research. In 2006, for example, Procter & Gamble Pharmaceuticals eliminated three hundred scientific jobs. "We have made a strategic choice to externally acquire rather than internally invent new medicines," a P&G spokesman was quoted as saying in *Chemical & Engineering News*. The same report noted: "The company argues that, with pharmaceutical compounds being developed by 4,400 biotech companies, many of which lack experience and funding, a partnership model is now the most efficient way to bring drugs to market." Other firms, including Merck and Johnson & Johnson, have adopted the same policy of cutting back on their own research in favor of buying the discoveries of entrepreneurial scientists, most of whom are professionally connected to academe.[2] For the multi-billion-dollar pharmaceutical companies, the buyout costs are cheap—typically $10 million to $50 million. For the academic entrepreneurs who have nurtured their start-ups from a glimmer of an idea to a patentable product, the payoffs bring unbelievable personal wealth. For the universities that owned the patents or held equity in the start-ups, welcome millions suddenly appeared.

The evolution and growth of this industrial sequence explains why tech transfer and entrepreneurship now figure in academic administration, planning, and allocation of scarce money, though not at every place in academe, since cultural unease regarding science for sale is far from dissipated. But the broad trend is toward commercialization. Universities with a head start in business enterprise strive to maintain their lead, while the laggards try harder to improve their performance. Here is a typical case of the latter: With marching orders to catch up, Richard Bruno, a successful high-tech entrepreneur and computer-science academic, was appointed director of McGill University's Office of Technology Transfer in 2004. Canada, he explained to me, was ten to fifteen years behind the United States in linking academe to industry. "We have to get the faculty to think differently," he said. "The core problem is they don't know, have never been exposed to business." Of fifteen hundred principal investigators at McGill, Bruno said, only thirty to forty were engaged in technology transfer. Getting the faculty's attention to educate its members about technology transfer was his most urgent and challenging task, Bruno said—a sentiment heard even at some universities long involved in technology transfer.[3]

Academic commercialism gets mixed reviews. Economists point out that the cash rewards are unevenly distributed among the many participating schools, that blockbuster deals are rare and generally confined

to a few major research universities. A chorus of dismay hammers on
the theme that money-chasing has gone too far, that without restraints,
academic science and commerce make an unwholesome mix. The case
against is buttressed by the sorry examples of mercenary scientists misusing their academic positions and the prestige of their institutions to
pursue commercial goals and enrich themselves along the way. Sinners
in science provide the critics with ammunition for attacking the linkages of academic research and commerce as inherently inimical to
traditional scientific values. Nonetheless, attitudes have changed. The
abuses that have arisen from the intimacy of science and industry are
condemned all around—though with varying degrees of sincerity and
determination to abolish them. But a lot of steam has gone out of the
belief that the linkage of universities and business is fundamentally unholy. I found that the convergence of the two sectors is viewed approv-
ingly by some elder statesmen of science who once observed it warily
and with misgivings. Bruce Alberts, an alumnus of the go-go days of
biotech at the University of California, San Francisco, told me how his
own attitude toward scientific entrepreneurship underwent a striking
reversal:

> I came to UCSF in 1976 just at the time when the first biotech
> companies were forming. And my friend Herb Boyer was talk-
> ing to me—did I want to do this, be on his board or some-
> thing like that, get involved with it? My whole reaction at that
> point—having come from Princeton, ten years in an academic
> environment, before that at Harvard—was basically, this was
> a world I didn't want to get involved with. It didn't seem either
> interesting or consistent with what I wanted to do. Any sense
> that you'd be doing something for money seemed to contra-
> dict the values of the community. And anyway, we were all so
> busy. All those things would be distractions from our main
> activities, and there was a sense that we would lose our reputa-
> tions among our colleagues if we showed we were going to be
> distracted from our main goal—*finding out the truth.* So, in
> retrospect, I saw this whole atmosphere change over the course
> of the seventeen years I was at UCSF. And I think, in retro-
> spect, people like me were wrong to be such purists, because
> I think there's a great deal of synergism that can be derived
> from thinking about an application of your work to practi-
> cal problems. I think our attitude was that we were doing our
> jobs so long as we published our paper, and that was serving

the American public. Since that time, I would take a different position. I think the academic community has an obligation to try to make sure that what we do—we're heavily supported by public funds—really does have a payoff, and that we could do a lot to speed that process if we didn't have this purist attitude that we're never going to walk inside a company and talk to them, much less start a company. I think a lot of great science is being done now in these biotech companies in ways that can teach the university more about how you can do more collaborative work.[4]

Harold Varmus, also at UCSF in that period, says he underwent a similar conversion regarding industrial relations with academe. "It's very hard for me to say that life hasn't changed," he told me, recalling that

before Bayh-Dole, I was a "lab rat." I didn't know anything outside my own lab and my own department. I was much more anxious about the potential damage that would be done. Not by patenting itself. I was more concerned about drug companies coming in and buying my colleagues. That worried me. But I've lived in this world now for a long time, and I've seen some of my colleagues get very rich. Even I made a little bit of money—a small nest egg. And I'd say I think the biotech industry has been good for the country, good for the world. And it's enriched our science. I think it would be very difficult to have the successes the biotech industry has without somebody making something.

I asked Varmus whether he was concerned about academic basic researchers shifting to applied research. He responded that if the shift arises from their own interests, rather than commercial inducements, "and they become more applied in orientation, that isn't necessarily a bad thing," adding:

My own lab suddenly evolved for the first time something which is pretty practical. It doesn't have financial implications. We're working with genetic changes in human lung-cancer patients. It has very clear diagnostic and therapeutic implications. We're actually not making drugs or making anything for sale. But we are changing the way medicine is practiced. And I actually find it exhilarating. It's a huge health problem, and we can make a contribution to it.[5]

As I often encountered among scientists whose careers spanned the old and new worlds of science and commerce, Varmus tempered his approval of the new relationships with scorn for the excesses that have accompanied them. "One of the things I really resent is the idea that everybody who discovers a mutation of certain genes can patent that mutation. That's absurd," he said. "You end up with patent clutter— too many patents on the same thing. And then it has become unaffordable. The whole point is to make sure things get out there and get out there at a reasonable price." And then he brought up the Onco Mouse, the Harvard invention, licensed to DuPont, that, for many, symbolizes ungoverned scientific commercialism, to the detriment of science and the public it claims to serve:

"One of my longest and most acrimonious debates with industry has been over the Onco Mouse patent," Varmus said. "I would have preferred that it had never been issued. It has very measurably slowed down research on cancer by inhibiting the use of mouse models by industry. DuPont has acted in a highly unethical manner. They charge extortionist fees. They've created the need for all kinds of legal services that never should have been necessary. This institution," Varmus asserted, referring to Sloan-Kettering, "will never take out a license. We just do things illegally, in view of DuPont."

"Have they challenged you?" I asked.

"They've not come after us," he replied. "I'd welcome them to come after us. Actually," he said, with a wink, "I have a secret plot here to rile them up to come after us. I don't know if that ever will happen. 'Come sue me.' It hasn't happened yet, but I'd be happy to bring them out. I think public opinion would drown them."

Entrepreneurial U

Teaching, research, and variously defined commitments to public service have long characterized the major-league institutions of higher education. But in recent years, the trio of traditional roles has been joined in many schools by a hearty newcomer: entrepreneurship. Differently explained and promoted from one university to another, it usually includes the sale of the scientific skills and output of faculty members, along with the creation of start-up companies to develop their science into marketable goods. The growing cash value of scientific knowledge coupled with the financial insatiability of universities were natural inducements to the marketplace. But the movement is also propelled by evangelical efforts in behalf of academic capitalism, from within and around academic institutions.

Even if they weren't hushed oases of scholarly endeavor, proverbial ivory towers, universities once cultivated separation from their surrounding communities, which often led to town-and-gown frictions between reputedly snooty academics and plebeian neighbors. Today university public relations offices energetically produce news releases, slick brochures, and colorful magazines telling of the university's ambitions, growth and accomplishments, good citizenship, and, especially, neighborliness. The university as an entrepreneurial powerhouse, creator of jobs, and community benefactor is a frequent theme in this booster literature, embellished with colorful photos of collaborations among students, faculty, administrators, alumni, donors, and business and community leaders. "Ivory tower," formerly a droll metaphor for academic isolation or escapism, has been transformed into a pejorative. "No one mistakes Penn for an ivory tower. And no one ever will," Amy Gutmann declared at her inauguration as president of the University of Pennsylvania in 2004. Penn, she said, will "engage locally and globally." The federal, state, and local governments—in harmony with business interests, boosters, and private philanthropies—endorse and support academic entrepreneurship. Substantial sums are offered to launch and nurture academic-industrial enterprises, in "incubators" and "hatcheries," in subsidized research parks, and by furnishing venture capital for fledgling firms. "Follow the money" is a principle of modern academic management.

A major philanthropy, the Ewing Marion Kauffman Foundation—of Kansas City, Missouri, net assets $1.7 billion in 2004—is dedicated solely to promoting entrepreneurship, including university-based entrepreneurial activities nationwide. In early 2003, in furtherance of this interest, the foundation invited thirty universities to compete for five to seven awards of up to $5 million each in the foundation's newly established $25 million Kauffman Campuses Initiative. A press release from the foundation described the program as "the first such effort of its kind" and explained that the goal

> is to transform campus life so that entrepreneurship is as integral and natural a part of the college experience as dorms, cramming for exams, and parties. . . . The Kauffman Campuses Initiative builds on the Foundation's rich, 10-year history

of advancing entrepreneurship, which has included substantial
support of entrepreneurship education at hundreds of U.S.
colleges and universities. . . . "We hope to effect nothing less
than a cultural transformation," said Carl Schramm, president
and CEO of the Kauffman Foundation. "These grants will un-
leash the power of entrepreneurship on campus. We want *all*
students—not just those enrolled in business or engineering
schools—to have access to the skills, orientation and networks
that lead to greater opportunities for them and result in more
jobs, innovation and prosperity for America."[6]

There was a time when many academics would have found puzzle-
ment, if not grounds for alarm, in this lavishly ambitious undertaking
to render, of all things, entrepreneurship—never explicitly defined—
into a commonplace aspect of American higher education. However,
from Shakespeare to prison management, anything goes in modern
academe, including the study and practice of entrepreneurship in universities across the country. "The business of America is business,"
Calvin Coolidge declared in 1925. And in academic confirmation
of this presidential observation, business degrees predominate by
a wide margin in undergraduate education, accounting for 307,000
of the 1.4 million degrees awarded in 2004. (Education was a distant second, with 106,000.)[7] A survey titled "Endowed Positions
in Entrepreneurship and Related Fields in the United States" found
406 academic "chairs and professorships in entrepreneurship" in 2003,
an increase of 169 in four years. For the purpose of the survey, entrepreneurship and related fields were defined as "a set of disciplines
interested in the creation, management, and growth of firms in society."[8] Responses to Kauffman's invitation to apply for money were not
lacking.

From the original thirty universities invited by Kauffman, fifteen
were selected to receive $50,000 grants to develop detailed proposals
"for an innovative, comprehensive five-year plan to inject entrepreneurship into the fabric of the university." According to the foundation,
teams led by the president or chancellor of each competing university
presented their plans at a meeting at Kauffman headquarters, in Kansas City. With the aspirants located far and wide, the presentations
entailed travel, plus preparatory effort by customarily overburdened
academic chieftains for a shot at a relatively small return. But when
money is offered from any but patently disreputable sources (and in
some instances they are not excluded), universities show up—recall-

ing the technique of the Washington celebrity hostess Perle Mesta for drawing a crowd to a party: "Just hang a lamb chop in the window."

In December 2003, grants ranging between $2 million and $4.5 million were awarded to eight universities: the University of Rochester, Wake Forest University, Howard University, Florida International University, the University of Texas at El Paso, Washington University (St. Louis), the University of Illinois at Urbana-Champaign, and the University of North Carolina at Chapel Hill. In announcing the winners, the Kauffman Foundation noted that it had initially imposed a matching dollar requirement of two to one for the awards, a standard technique for expanding philanthropic impact by requiring the winners to raise additional funds. The announcement noted, however, that the "selected schools were so enthused about the concept, they succeeded in securing a three-to-one match. As a result, the initial $25 million commitment will be leveraged into a $100 million investment for the creation of new interdisciplinary education programs." [9]

The University of Rochester, which received $3.5 million from the foundation, apparently was so enthused that in May 2005 it awarded Kauffman CEO Carl Schramm the university's Eastman Medal, "which recognizes individuals who, through outstanding achievement and dedicated service, embody the high ideals for which the university stands." [10]

Following the Money

In science and medicine, as well as in other fields, Washington University, in St. Louis, ranks high among the nation's one hundred or so major universities. Founded in 1853, it is financially strong, proud, and self-confident but, unlike many of its academic peers, culturally conservative and undemonstrative, which accounts for its modest visibility on the national landscape. Wash U, as it's known, lists twenty-one Nobel laureates in physics, chemistry, medicine or physiology, and economics who, from 1927 to 1998, were on the faculty when they received the honor or who did at least part of their Nobel-winning research there before moving away. Fifteen faculty members were in the exclusive National Academy of Sciences in 2005. Wash U ranked twelfth among the nation's universities in the coveted *U.S. News & World Report* ratings. It was number thirteen in endowment holdings, with $4 billion in 2004, and nineteenth in receipt of federal research money, $371 million, plus another $118 million in research sponsored by industry and other sources, including its own funds.[11] It ranks with several of the most selective schools nationwide in admission of undergraduates. In conversations with me, several Wash U administrators likened their school to Stanford in terms of size, wealth, graduate programs, and academic quality. On the grim side, however, Wash U is located in an economically sunken city. The population of St. Louis declined by nearly 15 percent between 1990 and 2002, and the unemployment rate has exceeded the national average for many years.

In the closely watched tabulations of campus-based research spun into patents and, ultimately, cash income, Wash U's performance was actually better than most, relative to its research spending, with eighty-one licenses and options yielding $12.5 million in 2003. Duke University, with similar research spending, took in only $2.7 million in license income that year. But even with lower research budgets, some universities were doing far better than Wash U. Caltech, for example, with $367 million in sponsored research, earned over $25 million, while Michigan State, with $321 million in research, earned $24 million.[12] Big moneymaking from academic research is heavily subject to the rare chance of a faculty scientist producing a discovery that can be converted into a blockbuster product. But the situation appeared favorable for Wash U to increase the take from its laboratories and also deliver economic benefits to the surrounding community. Other universities have successfully focused their scientific prowess on creating businesses, jobs, and economic growth by orienting their curriculum, priorities, and resources toward that goal. Why not Wash U?

Even before the Kauffman Foundation announced its $25 million program to encourage academic entrepreneurship, Wash U had taken steps to encourage the entrepreneurial spirit among students and faculty, though not with the campus-wide gusto of the Kauffman enthusiasts. The earlier effort was financed by a series of grants, eventually totaling $5.3 million, from Robert Skandalaris, a Michigan businessman, and his wife, Julie. In 2004 Wash U established the Skandalaris Center for Entrepreneurial Studies—which became the base for expansion with $3 million awarded by Kauffman. A Kauffman Foundation report on Wash U's winning grant proposal summarized the university's plans as follows:

> We will: create a scholarship program expressly for students interested in entrepreneurship; increase from 1 to 7 the schools [already within the university] teaching entrepreneurship; plan to start 15 new courses; offer grants to motivate faculty to create more; engage approximately 20 faculty at the outset; reach

students at all degree levels; open new centers; offer an undergraduate 2nd major in entrepreneurship; fill a professorship;
and nearly double the number of student-run businesses.[13]

At Wash U the foundation's money has underwritten the Kauffman
Fellowship Program in Entrepreneurship, described as a "collaborative
effort" that "connects the Olin School of Business, the School of Medicine, the School of Law, the Graduate School of Arts and Sciences and
other parts of the university. . . . Through this program, Ph.D. students
in the Division of Biology and Biomedical Sciences have a unique opportunity to learn how scientific discoveries are translated into successful commercial ventures."[14] The Skandalaris Center, budgeted for $15
million over five years, reported in 2005, "The medical school is teaching bio-entrepreneurship and has appointed eight Kauffman research
fellows. The business school has modified its courses and curriculum
to welcome other schools' students including new cross-listed offerings
that promote student collaboration and interaction. The new art school
studio is operating and the new law school clinic is working with students and the community to support cross-campus learning."[15]

The chief in-house leader for this effort is Ken Harrington, managing director of the Skandalaris Center and lecturer in entrepreneurship
at Wash U's Olin School of Business. In a soft conversational manner
but with evident conviction, he related a broad vision of entrepreneurship, explaining to me that it "is not about starting a business. It's
something larger than that. And when you recognize that, it opens up
the campus to act in a more collaborative cross-campus way, because
there's space around the term 'entrepreneurship' for everybody." Harrington, a successful entrepreneur and corporate executive before join-
ing Wash U in 2002, says that the nation's "Founding Fathers were
entrepreneurs; Martin Luther was an entrepreneur; impressionistic
painters were entrepreneurs. And they had nothing to do with founding businesses." The award of the Kauffman money and the entrepreneurial enthusiasm of Wash U's chancellor, which Harrington deemed
crucial, "just released all sorts of constraints in terms of the culture. In
a year, we went from eight faculty involved in entrepreneurship to over
eighty."

At the medical school, Harrington said, eight PhD candidates,
"nominated by the heads of their labs as being more entrepreneur-
ially oriented students," have been appointed Kauffman Fellows, "and
we pay ten grand of their PhD stipend for a year." The fellows, he
explained,

take the bio-entrepreneurship course, and they look for things that may benefit from some translational research activities, which we define as that gap between basic scientific research and publishing your paper and having anybody willing to put money into it. So, there's this sort of hole in the middle. We targeted learning and resources and energy and incentives at better understanding as a community what that middle space is and what causes it. So we have computational biologists in the class and we have biochemists, we have a mechanical "nano" [nanotechnology] guy, we have radiology folks, MBAs, the IP [intellectual property] clinic people willing to support prior art patent searches for any ideas. We have an art school studio that will do a logo if you get a little further along.[16]

The transformation of Wash U into an entrepreneurial institution is also the heartfelt goal of Michael G. Douglas, associate vice chancellor and head of the university's Office of Technology Management. "Washington University has never really paid much attention to this sort of issue of commercialization," he explained to me as he described the background for the changes under way. "Again, it's a culture here that is very much unlike that culture that you would find, let's say, at a Stanford or MIT—very academic and not inclined to commercialization." An industrial entrepreneur heading his own start-up company, Douglas was recruited to Wash U by Chancellor Mark Wrighton, formerly the provost and a longtime professor of chemistry at MIT, home of the outstanding, pioneering MIT Entrepreneurship Center and a flock of other programs linking campus and industry. Wrighton, he recalled, "said he wanted to see a community built around Washington University." As explained to me by Douglas:

It would be basically a community that creates jobs for the students of this university and its graduates. Creates opportunities for it to recruit faculty to this university, creates opportunities to retain faculty for the university. So you sort of step back and you think about it for a minute. What Mark was talking about, what he continues to do by his actions, is to see if we can repeat in certain aspects what has happened in Boston with MIT. He's a former provost at MIT. So it's basically the MIT model that we're seeing. Wash U, however, was generating very few invention disclosures, and invention disclosures are what you need to start the whole commercial engine. If you look at where we compare to our peer institutions, back to the Stanford and MITs, we generate invention disclosures at about one-fourth the rate of our peer institutions. And if you look at what we generate in terms of disclosures on a per-research-dollar basis, it's low, and it's low because I attribute this, number one, to the lack of an entrepreneurial environment within the university. The point that I use in my public descriptions of this to faculty and presentations to faculty groups is you have an academic duty to see the value of your creations properly delivered for the public good. And how can you get up in the morning and look at yourself in the mirror and be comfortable with what you're looking at if you're not doing that? So, it's a guilt trip.[17]

Douglas added that disclosures of scientific inventions have gone up about twofold. "We're not there yet, and that reflects a number of things, but we're getting there."

In 2005 the Kauffman Foundation hung another lamb chop in the window, this one aimed at economists. In a full-page advertisement in the *Chronicle of Higher Education* headed "It doesn't have to be a dismal science. You could study entrepreneurship instead," the foundation invited economists to apply for funds to "deepen understanding of how entrepreneurial activity drives the economy." [18]

Persistent residues of academic piety preclude uninhibited commercialism, but the old taboos have faded under the need for money, the public-service appeal, and the envied successes of institutions that have prospered by aggressively engaging in entrepreneurial activity. The young grow up in the new academic-business environment, hearing stirring calls from high places on campus for them to join in. The concerns raised by critics who see a costly downside in science for sale receive public attention when a grisly episode occurs, such as the death of a volunteer in a financially high-stakes medical trial, or other news that invites doubts about the wholesomeness of contemporary scientific behavior. However, apart from these infrequent eruptions into public view, concerns about the social, economic, and scientific costs of the prevailing mode of academic-industrial collaboration are largely confined to obscure conferences and publications. Several recent books and articles exposing the depredations of the pharmaceutical industry have brought some of these issues to a wider public—especially by three former editors in chief of the *New England Journal of Medicine:* Marcia Angell, *The Truth about the Drug Companies: How They Deceive*

Us and What to Do about It (Random House, 2004); Jerome P. Kas-
sirer, *On the Take: How Medicine's Complicity with Big Business Can
Endanger Your Health* (Oxford University Press, 2004); and Arnold S.
Relman, author of numerous blasts at academic-pharmaceutical financial ties, in popular and professional publications. Also of note is the
Integrity in Science Project of the Center for Science in the Public Interest, a Washington-based public-interest organization that diligently
tracks financial links between corporate organizations and individual
academics and professional societies. But these well-informed efforts
have not evoked any evident groundswell of public reaction, nor have
they disrupted the onward anti-regulatory push of the George W. Bush
administration.

Variations within Academe

While all universities are similar in many respects, each varies according
to its own history, circumstances, surrounding influences, and current
leadership. The penetration of entrepreneurial goals and values, in whatever ways they are defined, is markedly uneven across academe. This is
true even in circumstances where external powers, such as state legislatures and business and industrial groups, pressure the university to produce tangible economic benefits. While some universities, like Wash U,
enthusiastically cultivate and grasp the opportunities for entrepreneurship, others grapple with inexperience in business dealings or vestiges
of cultural aloofness from commerce. To an extent that I found surprising, a persistent, under-recognized indifference, wariness, or even resistance to commercialism survives in the contemporary academic world.
Marvin G. Parnes, associate vice president for research at the University
of Michigan, Ann Arbor, told me about chairing a committee drawn
from various Michigan campuses for "developing a proposal on how to
improve our interface with industry in collaboration across the state in
some technology sectors, with the premise being that we have a lot of
assets that are not exploited." Parnes said that "the irony from my perspective is it's been difficult to—it's opened up a great deal, but it's not
like the culture here has been truly infused and swamped by corporate
perspectives. It's more that we've had to make a strong effort to build
those linkages and try and encourage some of those initiatives." [19]

Especially in times of economic stringency, legislatures and gover-
nors demand economic returns from their state-supported universities.
But fed by news accounts of universities turning science into cash, the
pressure persists even in good times. Robert Kelch, who became vice

president for medical affairs at the University of Michigan in 2003 after serving in a similar position at the University of Iowa, told me, "They do it in Lansing and they do in Des Moines," referring to the state capitals. "I've been asked very difficult questions: 'Why haven't you developed more companies?' 'How many jobs have you developed?' 'What have you done for me today?' I give them the results, which are going in the right direction, and tell them that these things don't happen overnight. But we are doing better and we will continue to do better. But, it's not going to be our core mission," he said, sounding a theme strikingly different from the entrepreneurial rah-rah at Wash U and other universities.[20]

One of them is Georgia Tech, where Charles L. Liotta, vice president for research and dean of graduate studies, enthusiastically endorses Bayh-Dole as a powerful potion for boosting economic development. "It was an ingenious thing they came up with," he told me. "Because just think about it: I make a discovery in my laboratory and I publish it. So it's there for everybody. What company is going to take that discovery and run with it when another company can come along and compete with it? So by protecting the IP [intellectual property], you give more assurance that it can come to fruition in terms of new jobs and new companies." With a curriculum vitae stacked with publications, Liotta heads a major chemistry laboratory at Georgia Tech, in addition to his administrative duties, and he is also a consultant to DuPont and other corporations. "And through my consulting, through the discoveries we make in the laboratory, through the students that I educate, I have contributed to economic development."

"Remember," Liotta asserted, "our major goal is not to make money," an assertion often repeated at universities, both entrepreneurial and passive, though, as we've seen, it is not uncommon for universities to engage in costly litigation to protect the income from challenged patents. "Our major goal," Liotta explained, "is to be a driving force for economic development in the state. And if we happen to make some revenue from licensing or equity in a particular company, that's good, because we put it back into research. But our prime goal, produce students, that's number one. Number two is to be on the forefront of research for economic development."

Liotta noted, however, that relations with industry can conflict with academic values:

> We're still learning how to do things. If any institution tells
> you they've mastered how to do this, they're not telling you the

truth or they're naive. The reason is whenever you get faculty and students involved in start-up companies, licensing, there are potential conflicts of interest that arise normally. To be able to do economic development at a university, you have to get your arms around how to manage potential conflicts of interest. You do it case by case. If anyone has a general rule for doing it, again, I think they're either naive or they're lying to you. I paint a very nice picture of the entrepreneurial spirit here, but along with that comes an obligation of managing all these potential conflicts that can arise. Can they be managed? Absolutely. We have no problem doing it. But, one, it has to be recognized and, two, you have to put in place an operational way of managing it. Whenever there are potential conflicts of interest that arise, I form a committee to follow it through.[21]

At Georgia Tech, as at many other universities, the linkage of campus labs and industrial opportunities is enthusiastically endorsed, starting at the top with the school's president. Considerable resources and constant exhortation are devoted to that goal. Entrepreneurship is in the atmosphere, proudly acknowledged. (In chapter 9, I'll converse at length with a young researcher at Georgia Tech who is deeply immersed in basic research while collaborating with a firm interested in commercial applications of his work.) Roger P. Webb, chairman of the School of Electrical and Computer Engineering, a major contact point with industry, told me in a matter-of-fact manner, "Georgia Tech is an entrepreneurial place. There has always been some aspect of that in the culture. It got heated up by the technology boom in the '90s. People saw a potential to make money."

"But what if a faculty member doesn't wish to engage in entrepreneurial activity?" I asked Webb. One might as well have asked how a pacifist would fit in at West Point. "Oh, yeah, there are faculty members who have no interest in that," Webb acknowledged, adding, "And there are faculty members for whom that's beneath their academic dignity. You just leave them alone," he said. "If they don't have the internal incentives to do that, they're not going to do it. They'd much rather produce learned papers for their peers. And that's fine." He noted, however, that twenty-six new faculty positions financed by a special state appropriation "were all recruited on the basis of being active in developmental activities."[22]

Managing a state institution in a relatively poor state, Georgia Tech's administrators unhesitatingly proclaim a responsibility to create jobs and wealth for Georgia's citizens. At the Manufacturing Research Center, a prominent, modernistic structure on the Georgia Tech campus, Director Steven Danyluk, professor of mechanical engineering, orchestrates the operations of sixteen spacious laboratories that exist for the purpose of working closely with industry. Fifty faculty members work in the center, along with 120 graduate students. As he explained to me:

> This particular center is unusual as far as university centers go. We get about 70 percent of all our funds from industry. So, we're always out recruiting industry members to come in here and do research with us. Typically in other universities, 20 percent is in industrial funding. We're sort of the opposite, in the sense that we're very used to doing business with industrial members, industrial companies. Companies put people in here, and I have offices in this building for company staffs. So, again, we're very used to interacting with companies. So the whole business of tech transfer is not something that is just occasionally done. But it's the *major* thing that we do. That's kind of how we survive in the research field.

Students at all levels—undergraduate, graduate, and postdoc—participate in the industrial projects, he pointed out. "One of the ways I put labs together is I put real industrial equipment into the labs, so students are actually working on real industrial-strength equipment. And that's a big advantage to industry, because they don't have to retrain students"—many of whom, following graduation, directly go to work for the firms conducting research at the center. "That's our number one product," he said. "We're a manufacturing center, but we don't manufacture anything but students. That's our product. And so fundamentally, the number one reason companies come to us is to hire students." "And coincidentally conduct a research project?" I asked. Danyluk replied:

> Every now and then they'll want something done, but generally speaking, if it's something critical to the companies, they can do it themselves much quicker and better than we can. And they also come to us because they will learn things that they don't normally do in their everyday world, or in areas totally outside of their normal business areas. So what they like to do is learn new processes, new techniques, new characterization

tools, whatever, that are being developed in other fields that they will then import into their field. And that's really what they like to do. A lot of time they will stimulate their own staff members' thinking by having them located here.[23]

The Lingo of the Market

Today's shoptalk within the scientific professoriate includes patents, licenses, contracts, intellectual property, venture capital, angel investors, industrial incubators, spin-offs and start-ups, equity, and other terms from the once-distant world of business and moneymaking. The term "burn rate" means nothing to most people. But among entrepreneurs, it's a clearly understood concept borrowing from rocket-fuel consumption and applied to the spending pace of capital on hand for a business enterprise—usually loans or investments for a start-up company. Consume the money too quickly and, like an out-of-fuel rocket, the company will crash. Few universities have embraced entrepreneurship in the whole-hog fashion celebrated and subsidized by the Kauffman Foundation. But many universities sponsor organized programs to encourage their scientists to be alert to the commercial potential that may exist unnoticed in their laboratories. Researchers are cautioned that failure to make the disclosure required by the Bayh-Dole Act in a timely fashion can put a discovery into the public domain, ineligible for patent protection and up for grabs by anyone.

Careers sometimes oscillate full-time between campus and the corporate sector, with returnees to academe linked to industry by board memberships. John L. Hennessy, an electrical engineer and computer scientist, joined the Stanford faculty in 1977. In 1984, during a sabbatical year, he cofounded MIPS Computer Systems, Inc., to commercialize his research in microprocessors. After returning to Stanford, he was appointed to an endowed chair in 1987, dean of engineering in 1996, provost in 1999, and president of the university in 2000. Hennessy serves on the boards of several corporations, including the high-flying Google, Inc., which accompanied his board membership with sixty-five thousand shares of company stock. At the University of California, San Francisco, Joel Kirschbaum, director of the UCSF Office of Technology Management, also runs his own private biotechnology consulting practice. Kirschbaum explained to me that he operates the business outside of UCSF office hours and told me: "I do not consult in my private practice on any matter relating to UCSF, its investigators, or its technologies, nor would I conduct business on behalf of UCSF with a company in which I had a financial interest (but would refer that project to someone else at UC to handle instead)."[24] In and around academe, the ongoing commercialization of science is widely hailed as a welcome development. The benefits are calculated in counts of inventions, patents, licensing, and the revenues they produce, and are also seen in the growing multitudes of personal and business relationships linking campus and corporation. Reflecting on a bygone time, David Korn, the former Stanford medical dean, observed that

> the research universities are moving away from an era where reward was nonmaterial, that is, you wanted to be best in this, you wanted Nobel prizes, academy membership of your faculty, prizes. The faculty wanted to be the first out with the paper, beat the competition, be recognized for a fine piece of scientific work, be invited to be a plenary speaker at a prestigious meeting, maybe win some prizes. That's what drove my colleagues and me when I was running a lab. I didn't do experiments because I thought I was going to get products, get money, and this and that. The reward system was geared to nonmaterial things. And I think that to an extent that does trouble me and maybe others a lot, that the reward system has swung toward material reward.[25]

5

The Price of Profits

Universities will tell you simultaneously that this kind of dealing with industry is a whole new way of doing business and has altered the nature of
the university and changed the research agenda. And then if you're talking
to them about whether it is having any effect, you'll hear that universities
are the same as they were in Bologna in 1500 and nothing has ever happened
and this doesn't mean a thing. Presumably, both of those are not actually
true. But it is extraordinarily hard to have a discussion about the effect
that university-industry cooperation has on either conflicts of interest but
also on broader issues in terms of the nature of faculty-student relationships, the nature of faculty-faculty relationships, the nature of the research
agenda, the nature of publication. And what I fear is this is parallel to what
happened when, in the post–World War II era, research dollars started pour-
ing in, no one ever wanted to have a discussion about what impact that was
having on the teaching side of the university. Everyone just wanted to say
there was no change whatsoever, even as the university was fundamentally
transformed.

David Goldston, chief of staff, Committee on Science,
U.S. House of Representatives[1]

: : :

Entrepreneurship can easily conflict with the idealized
concept of science as a noble, public-spirited enterprise,
desirably different and separate from material matters.
And yet, in the post-Bayh-Dole era, law, public policy,
and the potential for moneymaking and job creation
have pushed academic science toward intimacy with

entrepreneurship. Their embrace is encouraged by the federal and state governments, university, business, and community leaders, and is warmly treated by the popular press, with rarely a skeptical note. The movement toward academic commercialism has raised danger warnings, such as those expressed by former Harvard president Derek Bok and others who see incompatibilities between marketplace mores and integrity, collegiality, and other valued attributes of the traditional scientific culture. These relatively few critics, however, are scarcely heard amid demands for even closer academic-corporate ties. The movement rolls on—beneficially so, say its supporters, who often scoff that science is being a bit precious in its fretting about the risks of commercial contamination.

As in other activities, when big money flows fast, temptations and opportunities arise for risky behavior and stealthy or even brazen wrongdoing in pursuit of personal or institutional advantage. The new world of academic-commercial dealings is characterized by some gray areas and evolving rules for permissible and impermissible conduct. The people who manage and conduct research in scientific organizations are not immune to the weaknesses and foibles so plentiful elsewhere, despite the accolades for probity that science bestows upon itself. Science possesses its share of dedicated scoundrels and careless wanderers over the ethical terrain. Transparency, openness, and disclosure are the most frequently prescribed measures for bolstering scientific integrity. But they are not always present, and when they are, they do not always suffice. Major deals between universities and corporate sponsors, on the scale of the Stanford climate-change study financed by Exxon-Mobil and other firms, are too big to escape notice and interest in their contractual details. In contrast, the details of innumerable commercial deals of a lesser scale usually receive little exposure to public scrutiny beyond a semi-opaque press release, and sometimes not even that. Many of these dealings between science and industry are characterized more by silence or murkiness than by the vaunted transparency that is celebrated as a guarantor of upright behavior. It's usually only when something goes very wrong that the tawdry details of academic-industrial deals become visible to outsiders. The classic case occurred at the University of Pennsylvania in 1999 and has since reverberated as a metaphor for academic science gone off its ethical tracks. In the Penn case, a volatile brew of medical research and profit-seeking led to the death of an experimental volunteer, grave institutional embarrassment, and a blighted career.

The Scandal at Penn

Penn's Institute for Human Gene Therapy was a leading contender in a crowded race to exploit the revolution in gene research and cure untreatable maladies. Success would fulfill the humanitarian hopes of science and also, possibly, provide huge financial rewards for investors, researchers, and other stakeholders in the quest. Relatively few drugs that pharmaceutical and biotech firms enter into the FDA's regulatory gantlet emerge successfully, but some of those hit the jackpot, therapeutically and financially, bringing in hundreds of millions or even billions of dollars in sales that more than repay the costs of less successful and failed drugs. For universities that own the patents that underlie successful drugs, royalties on licenses to pharmaceutical firms can produce huge returns. In the 1990s, gene therapy was seen as the next frontier in medicinal innovation and profits. In accompaniment to this vision, the good-news apparatus of academic science and the biotech and pharmaceutical industries issued tantalizing reports of promising developments, though often based on fragmentary evidence derived from brief clinical trials involving small numbers of patients—sometimes below the count for statistical significance.

Backed by the NIH's billions and rising pools of private venture capital, the gold rush was unstoppable. In 1993, to lead its efforts in this promising field of research, the University of Pennsylvania recruited a top-flight gene-therapy researcher from the University of Michigan, James M. Wilson, who, in addition to his academic role, was a founder and 30 percent owner of Genova, Inc., a biotechnology company on the forefront of gene-therapy research. Big money was on the table, with the scent of much more to come. Genova agreed to provide Penn's Institute for Human Gene Therapy, headed by Genova stock-owner Wilson, with $4 million a year for five years, and to give the university a 5 percent stake in the company. The university agreed to give Genova exclusive licenses for developing drugs and treatments based on discoveries made at the institute, in effect establishing a monopoly from which Penn, Genova, and its stockholders would share the profits, if any. The freewheeling deal fit well with the entrepreneurial style of Penn president Judith Rodin, who took office a year after Wilson's arrival at the university. Rodin, building on efforts already under way when she arrived, drove hard to restore the academic distinction that had drained away from Penn, rebuild its finances, and stimulate commercial and residential renewal in Penn's blighted, crime-ridden surroundings on the edge of downtown Philadelphia. Investing in community housing, security, retail facilities, landscaping, and public schooling, Penn turned the neighborhood around and also regained its luster as an outstanding university. Visiting delegations from other universities sought lessons from Penn's rebirth. A coveted accolade came in 2005, when *U.S. News & World Report* listed Penn fourth in national college rankings; five years earlier, it was tied for seventh with Duke and Johns Hopkins.

In their incestuous dealings on the frontier of gene therapy, Penn and Wilson held a pecuniary interest in achieving success—generally regarded even then as a danger sign in the conduct of clinical research. On the principle that the safety and efficacy of a drug should not be judged by anyone who stands to profit from a favorable assessment, many institutions prudently prohibited clinical testing by scientists with a financial stake in the outcome. Situations in which the university itself owned a stake in a drug undergoing tests on campus proved to be more difficult to regulate because of complex financial entanglements and the potential for enriching the university treasury. But then, and still today, the comforting sense of goodness that pervades academic management assuaged concerns about institutional conflicts of interest. Penn should have recognized that its deal with Wilson and Genova was loaded with potential trouble. But as a private, independent institution, it was free to deal as it pleased within very broad limits and without public disclosure of financial details that might arouse critical notice. And so it did, even as pressures increased for universities to clean up their increasingly complex entanglements of research and profit-seeking, especially where patient care and safety were involved.

A Young Volunteer

In September 1999, into the conflicted setting at Penn came Jesse Gelsinger, age eighteen, diagnosed with a mild form of a rare metabolic liver disorder, ornithine transcarbamylase deficiency. Diet and drugs kept Gelsinger's health problem under control. Nonetheless, the youth volunteered to participate in a trial of a genetically engineered virus in the belief that yet unborn infants with the disorder might someday benefit from the research, though it would not help his condition. The FDA later concluded that Gelsinger was not told that several human patients had experienced serious side effects from the virus treatment and that deaths had occurred among several monkeys treated with the experimental virus. The informed consent form that he signed included only a brief reference to Wilson's financial stake in the drug trial. Shortly after Gelsinger received the experimental treatment, he developed a high fever. His organ functions rapidly deteriorated, and he died four days later. While offering condolences and praising young Gelsinger's altruism, Penn initially adopted a hard-nosed stance, insisting that Wilson, his collaborators, and the university had adhered to all relevant federal regulations and ethical standards. The university later conceded, however, that it had fallen short in meeting government reporting standards and in halting the gene trials and notifying the Food and Drug Administration of side effects among experimental subjects before Gelsinger received the fatal treatment. Following an investigation, the FDA took a harsh view of the situation, stating in a letter to Wilson that "you repeatedly and deliberately violated federal regulations in your capacity as investigator in clinical trials." [2] In 2000, a year after Gelsinger's death, the university entered into an out-of-court settlement with his family for an undisclosed sum, a common legal tactic for avoiding the expense and exposure of a public trial. In that same year, Genova was acquired by Targeted Genetics Corporation in an $89.9 million buyout that reportedly netted $13.5 million for Wilson and $1.4 million for the University of Pennsylvania. Regarding the sale, Bob Taber, vice chancellor of science and technology development at Duke University, was quoted in the *Wall Street Journal* as saying: "I suspect that this is a deal that officials at Penn are happy with because all the PR problems with the company will go away, and they're liquidating their investment at a reasonable value as well." [3]

In 2005, without admitting wrongdoing, the university and a collaborating institution in the disastrous experiment, the Children's National Medical Center in Washington, agreed to a civil settlement with the U.S. Department of Justice in which each paid fines of approximately $515,000. Wilson gave up human studies following Gelsinger's death, was barred from taking part in human experimentation until 2010, and was required to undergo training in the rules of human trials. [4] In academic-scientific circles, little sympathy was evident for Penn and its brazen, convoluted financial arrangements on the frontiers of gene therapy—partially from concern that the episode tarred all academic science and all academic-commercial deals. During a break at a congressional hearing on the Gelsinger case, I heard a Penn official grimly lament Penn's notoriety: "We're the place that killed the kid."

David Korn, whose wife was a vice president at Penn, told me that his knowledge and opinions of the Gelsinger case are based on the public record and discussions in biomedical circles, including the

Association of American Medical Colleges, where he's a senior vice president. Referring to his former position as dean of the Stanford University School of Medicine, Korn said:

> When I first heard about the deal that Penn had offered Wilson, I had grave concerns about it, because we would not have done that at Stanford. We simply would not have allowed that. I actually faced down a chairman of the board of trustees at Stanford who tried a deal like that on one of our very research-intensive divisions of child psychiatry. I turned it down and he appealed to the president, that was Don Kennedy, and Don stood by me and told the chairman of the board that we would just not allow that to happen. I worried that they [Penn] were setting themselves up for something—for at least the perception of a disaster, even if no disaster occurred. But if anything occurred, the nature of that arrangement could only lead to withering criticism.[5]

A Permissive Atmosphere

Penn's headlong pursuit of scientific profits, cure, and glory led to the sorrowful outcome. But also in the picture was Washington's long-standing timidity in enforcing ethical standards in the universities thriving on government research money. The universities themselves, through the associations that represented them in Washington, expressed wariness of externally imposed strict rules and regulations. In the late 1980s, the Public Health Service, which included the NIH and the FDA, rumbled with concerns about conflicts of interest that might be incurred in the conduct of government-financed research. The PHS agencies directed grantee universities to require disclosure of outside income by their researchers, but the feds themselves maintained a respectful, arm's-length distance from their academic beneficiaries. The information remained on campus and was not subject to examination by government officials. An abundance of federal rules existed for the protection of human volunteers in medical experiments, but adherence was largely entrusted to an honor system. Medical researchers, and the popular culture surrounding them, regarded science as a humanitarian enterprise, bound by high professional standards, with no need of cops on the beat to ensure honest dealings. The federal agencies that might have intervened shied away from confrontation.

Within biomedical circles, however, awareness arose of ethically risky behavior on the increasingly commercialized frontiers of science.

In 1990—an ancient time in the evolution of science for sale—concerns about conflicts of interest in university-based biomedical research inspired the issuance of recommended "guidelines" by the Association of American Medical Colleges (AAMC), a professional league of the nation's then 125 mainstream medical schools and hundreds of teaching hospitals. Penn was a member but was undeterred by the guidelines, though they were mild, reflecting the association's preference to move gingerly. As a voluntary organization, it possessed no enforcement powers over its large and diversified membership, which ranges from world-leading, research-intensive medical schools to small, little-known state institutions. Basically, the AAMC is a service organization for its members, performing a critical role in American medical education: preparation and administration of the admissions test for the member schools, for which it currently charges $210 per applicant, of whom there are currently about thirty-seven thousand per year. It also provides other testing and administrative services for medical education, which bring in a total of over $40 million a year. Submerging differences, respecting local sensitivities, and holding the diversified lot together is the essence of association politics.

But headquartered in Washington, D.C., the AAMC was positioned to tune into the political chatter concerning the money for research and training that the NIH supplied to the association's member schools. The NIH bankroll on which they depended steadily grew in response to public hopes for cures and congressional pride in supporting appropriations for the renowned NIH. But with dubious dealings coming to light here and there in the new era of Bayh-Dole commercialization, the AAMC's Capitol Hill observers warned of a political backlash. As stated in the introduction to the guidelines, "Ruptures of public confidence that occur when biomedical researchers are involved in conflict situations, deliberate or otherwise, inflict long-term damage on societal trust and support." Nonetheless, the guidelines were tepid and permissive, yielding to the onward rush into commercial deals by AAMC member schools. Thus, the introduction delicately stated that "participation in a situation with opportunity for personal gain does not constitute an unacceptable situation of itself; it is the potential stimulus for unacceptable behavior that must be addressed"—Aesopian language to the uninitiated. But the motivation for the AAMC report and its message were clear for those who were aware of the difficulties and conflicts arising from the spread of commercialism into academic science. Contending that conflicts of interest were unavoidable in the new era of academic-commercial collaboration, and that not all were harmful, the AAMC guidelines cautioned that "conflicts become detrimental when the potential rewards, financial or otherwise, cause deviation from absolute objectivity in the design, interpretation, and publication of research activities, or in other academic and professional decisions." It added that "individual conflicts of interest arise in large part because of the interplay between a faculty member's personal and financial interests and the opportunity to conduct externally-funded research." Moving on, the AAMC guidelines presciently warned of the type of deal that Penn struck with Wilson and Genova, Inc. Under the heading "Situations That May Impart <u>Bias in Research</u>" (original underlining), it disapprovingly cited: "Undertaking basic or clinical research when the investigator or the investigator's immediate family has a financial, managerial, or ownership interest in the sponsoring company or in the company producing the drug/device under evaluation." [6] The AAMC did not recommend the prohibition of entanglements of money, self-interest, and research, but rather their disclosure, supervision, and management. In its deal with Wilson and Genova, Penn paid scant attention to that prescription, mild as it was.

The Stain on the NIH

The revolution in biotechnology was in large part financed by grants that the National Institutes of Health awarded to university-based scientists. It was furthered, too, by scientists in the NIH's own laboratories. NIH scientists are career civil servants or commissioned officers in the Public Health Service, with many visiting foreign scientists working at their side. The great lure of the NIH is the favorable conditions that it provides for its staff researchers: well-equipped laboratories, outstanding colleagues, generous financial and technical support for research, and freedom from teaching duties and the recurring grant competitions that make professional life harrowing for university scientists. NIH researchers are not exempt from the competitive rigors of modern science. They are periodically reviewed for the quality of their work, usually by scientists drawn from universities, but compared to the stressed environment of university-based science, the NIH is a comfortable place to work. Scientists interested in attaining great personal riches have no reason to be drawn to government service, where top salaries are generally linked to the pay that Congress votes for itself, $162,100 in 2005 for the great majority of members, a bit more for leadership positions. At the NIH, as elsewhere in government, sought-after specialists, including physicians and scientists, are eligible for supplemental pay, but big income and NIH employment did not go together—or so it seemed.

Even after the passage of the Bayh-Dole Act in 1980 and other legislation designed to encourage commercialization of government-financed research, the NIH remained largely untouched by the tides of materialism that were beginning to sweep over academic science. As discussed earlier, even today the NIH remains a laggard in fulfilling Bayh-Dole's toothless requirement for all government-financed researchers to assist in commercializing their scientific findings. But not even the NIH proved impervious to the lure of money, though only a very small portion of the research staff was implicated in the scandal that tarnished the great institution in 2003, when staff members' moonlighting deals with drug firms were exposed to the public. The sense of precious values betrayed was heightened by the sterling reputation of the NIH, by the unexamined myth that the great government institution was an oasis of selfless integrity, dedicated to science and healing, aloof from the new world of stock options, start-up companies, and lucrative consultancies increasingly common in academic science. Public scorn, professional embarrassment, and congressional ire resulted from the revelations of business dealings between several dozen senior managers and scientists at the revered NIH and the pharmaceutical industry. In the public mind, the industry was ambivalently regarded, renowned for its miracle products, but reviled for prices that led to news accounts of low-income retirees pathetically forced to choose between food and medicines. It was not a good time for NIH managers and scientists to be discovered in profit-making trysts with the pharmaceutical industry.

For openness and public accountability, the National Institutes of Health once appeared flawlessly transparent. Controversies would occasionally arise about the integrity of this or that scientist on its staff, or the accuracy of a publication, notably so in the marathon controversy in the 1980s between NIH researcher Robert Gallo and scientists at France's Institut Pasteur over priority for identifying the HIV/AIDS virus. But, with rare exception, the seemingly open-window view of the NIH confirmed the expectation of adherence to the highest ethical standards. The NIH annually reports in public session to appropriations subcommittees in the House and Senate and to other congressional committees as issues of particular interest arise. Unlike some other government agencies, the NIH dutifully responds to requests filed under the federal Freedom of Information Act, though that "sunshine" law is riddled with allowable exceptions that provide excuses for noncompliance. Meetings of the NIH's influential advisory committees are usually open to the public and are well attended by the press, by so-called patient advocates, and representatives of scientific and medical organizations. The NIH's press offices keep the news media abreast of accomplishments by NIH scientists and by NIH grantees in universities and medical centers. And its administrators and scientists have usually been easily accessible to the press, though less so under the restrictive information policies of the George W. Bush administration. The NIH has long been hailed as the jewel in the scientific crown of the U.S. government, lauded during the Reagan administration by the secretary of Health and Human Services, its parent agency, as "an island of objective and pristine research, untainted by the influences of commercialism." [7]

In the working assumptions of the popular and scientific and medical press that report on the NIH's work, scientific excellence at the NIH was a bedrock fact, and dedication to the public well-being was assumed without question. Nonetheless, even with the multiple observation sites available for journalistic and other observers, an infestation of dubious financial dealings between NIH staff members and pharmaceutical firms flourished without public knowledge from at least 1995 until 2003. It was painfully exposed through the investigative reporting of a single journalist, and it set off an earthquake in biomedical and related political circles. The episode illustrates the familiar risks of journalists abandoning skepticism and lapsing into symbiotic comfort with their subjects. With the NIH confidently regarded as an icon of scientific integrity and public accountability, why even consider the possibility of wrongdoing or give credence to whispers of impropriety? None but that single journalist came equipped with the needed skepticism.

Writing in the *Los Angeles Times* in December 2003, David Willman reported consulting fees in the hundreds of thousands of dollars and other income paid to some senior NIH administrators by pharmaceutical and biotechnology firms with clear interests in the NIH's research, collaboration, and good name. Opportunities for personal enrichment had invaded the cloistered calm of the NIH. In most instances, Willman reported, disclosure of outside income, commonly required for federal employees, was ignored, obscured, or hidden under a relaxation of conflict-of-interest and disclosure regulations that was intended to make the NIH competitive with the salaries and moonlighting opportunities of university employment. In some instances, consulting deals were not disclosed, in violation of NIH rules. When they were disclosed, the required paperwork was indifferently filled out and reviewed, according to a report by the inspector general of the Department of Health and Human Services. The report noted that at most of the NIH's twenty-seven institutes and centers, ethical monitoring was a "collateral duty," meaning it held a low priority.[8] Reflecting the NIH's lack of attention to outside dealings by its staff members was uncertainty about the numbers involved in these dealings. NIH director Elias Zerhouni responded to Willman's articles by appointing a "Blue Ribbon Panel on Conflict of Interest Policies," co-chaired by two pillars of the Washington science-technology establishment, Bruce Alberts, president of the National Academy of Sciences, and Norman R. Augustine, retired CEO of Lockheed Martin. Following a crammed two months of hearings, collection of information and comments via the Internet, interviews and writing, the Alberts-Augustine panel delivered its report to Zerhouni in May 2004—a high-speed performance relative to the scientific community's normally languid conduct and write-ups of studies and inquiries. Unwittingly, the report contained seeds of further difficulties for the NIH in stating that the investigators were "surprised to learn that relatively few NIH employees are in fact engaged in consulting agreements with biotechnology or pharmaceutical firms—an activity that involves only about 120 of NIH's 17,500 employees." The latter figure included all NIH employees, from clerks and scientists up to director.[9] As thus stated, the number of consultants appeared to be a minuscule portion of the NIH workforce, though, in fact the consultants were drawn from a much smaller pool, the 6,000 researchers and administrators among the 17,500 staff members. The panel accompanied this finding with recommendations for severe restrictions on outside commercial activities by senior NIH management, but the low count of consultants held the potential for reducing the crisis to the level of a few bad apples in a very big barrel. Worse numbers, however, soon surfaced, in embarrassing circumstances.

Congressional Guardians

Protectively hovering over the NIH in good and bad times was the House Committee on Energy and Commerce, chaired by Rep. W. J. "Billy" Tauzin (R-Louisiana), which held authority to write the laws for the NIH and scrutinize its operations. Chairing the committee's Oversight and Investigations Subcommittee was Rep. Jim Greenwood (R-Pennsylvania). In tandem, the two chairmen applied political heat to Director Zerhouni to root out illicit commercialism at the great research institution for which they held legislative responsibility. But then, as the congressional inquiry proceeded, along came a pair of developments beyond the plots of febrile conspiracy dreamers or script-writing specialists in surprise endings. In July 2004 Rep. Greenwood announced he would leave Congress to become president of the Bio-technology Industry Organization (BIO), the Washington-based trade association of mainly start-up and young firms that regularly draw science and technical guidance from consulting scientists in universi-ties and other nonprofit institutions. In its frequent representations to Congress, BIO insistently argued against restrictions on consulting by both academic and NIH scientists, for the very good reason that it de-pended heavily on their scientific knowledge and skills. With most of its member companies running on fast-burning venture capital, access to pioneering science—virtually all financed by the taxpayers—was a priority for BIO.

Then came a second surprise. First elected to Congress in 1980, Rep. Tauzin announced in February 2004 that he would not run for reelection. In December came the announcement that he would be-come president of the Pharmaceutical Research and Manufacturers of America, the multimillion-dollar Washington-based lobby for the major research-oriented drug-manufacturing companies. Big Pharma firms, as they are often referred to, regularly papered Capitol Hill with campaign money, providing $91,500 for Tauzin and $54,575 for Green-wood in 2002. Between 1998 and 2002, BIO spent $3.5 million on lob-bying, while political action committees of individual firms put out a total of over $30 million.[10] The congressional investigation of financial wrongdoing at the NIH thus produced its first benefits: from the merely comfortably paid ranks of congressmen, the two chairmen in charge of investigating financial wrongdoing involving NIH scientists and drug firms vaulted into the plush realm of major-league Washington lobby-ing for the organizations that represent the biotech and pharmaceuti-cal firms. Tauzin's Big Pharma salary was reported to be $2 million.[11]

Succeeding Tauzin as chairman was Rep. Joe Barton (R-Texas), and taking Greenwood's place was Rep. Ed Whitfield (R-Kentucky). While Zerhouni and company hoped the Alberts-Augustine report would douse interest in the commercial misadventures at the NIH, the two congressmen took a new, roundabout approach to the investigation. Rather than relying on the NIH to supply the names of staff members with questionable outside dealings, the committee shrewdly asked a score of drug firms to identify their NIH consultants. In response, the committee received eighty-one names that had not been included in the NIH's own count. The diligence of the NIH's own internal inquiry was

thus brought into question, further embarrassing Zerhouni and under-
mining his efforts to assure the NIH's congressional guardians that
the ethical problems on the Bethesda campus had been identified, were
relatively small, and were under control. Referring to the revelation of
additional names, the beleaguered NIH chief was quoted in *Science*
as saying, "It was like getting shot in the back by your own troops." [12]
Investigating the outside dealings of the previously undisclosed names,
the NIH said it found that thirty-six of these employees had "violated
policies or regulations and were referred for administrative action. In
addition, eight reviews found violations of policies or regulations by
individuals who are no longer NIH employees, and are not subject
to administrative action by NIH." Zerhouni noted in his response to
Chairman Barton an especially damning finding: that for some of the
companies engaged in financial dealings with NIH employees, "the
main benefit was the ability of the employer to invoke the good name
of NIH as an affiliation." [13] In the armory of marketing machinations
used by pharmaceutical manufacturers to gain the confidence of physicians who prescribe drugs, the good name of the NIH is greatly valued.
The NIH was trustworthy because it stood apart from the shameful
profit-chasing that infested science—or so it seemed.

Congressman Barton's discovery of previously undisclosed scientist-
consultants at the NIH, coming as it did after the Blue Ribbon panel's
comforting low count of those involved, reignited concerns about what
was going on out there in Bethesda. Bruce Alberts, co-chairman of the
panel, told me that

> the whole committee felt that we had been done in by these
> scientists, who obviously knew by the time our committee
> started, if they didn't know before, that they had to fill out
> those forms. And if they were honestly making a mistake, they
> should have filed these forms much earlier than being discov-
> ered by a congressional committee. So I have no sympathy at
> all for these scientists. It was very clear that they needed to get
> permission [to serve as consultants]. The rules were very clear.
> And if they didn't know about them, they certainly would have
> known about them after all the flak. I think we've all been
> betrayed, in a sense, by their misbehavior. [14]

Even before additional names surfaced, the reports of private dealings by NIH executives aroused concern and even revulsion in the biomedical research community. Among the NIH's high-level recipients of

outside income, Willman reported, was Stephen I. Katz, director of the National Institute of Arthritis and Musculoskeletal and Skin Diseases, who received company fees in the range of $476,369 and $616,365 over the previous decade. One company, from which Katz received at least $170,000 in fees, was awarded $1.7 million in grants by his institute before going bankrupt, according to Willman. Another NIH official, Ronald N. Germain, deputy director of a laboratory at the National Institute of Allergy and Infectious Diseases, received over $1.4 million in consulting fees over eleven years, plus stock options. Others received lesser but still substantial sums from pharmaceutical firms.

Willman's articles described numerous instances of NIH scientists and administrators collaborating in the corporate exploitation of the NIH's prized reputation—and along the way accepting large consulting fees:

> Dr. P. Trey Sunderland III, a senior psychiatric researcher, took $508,050 in fees and related income from Pfizer Inc. at the same time that he collaborated with Pfizer—in his government capacity—in studying patients with Alzheimer's disease. Without declaring his affiliation with the company, Sunderland endorsed the use of an Alzheimer's drug marketed by Pfizer during a nationally televised presentation at the NIH in 2003.
>
> Dr. Harvey G. Klein, the NIH's top blood transfusion expert, accepted $240,200 in fees and 76,000 stock options over the last five years from companies developing blood-related products. During the same period, he wrote or spoke out about the usefulness of such products without publicly declaring his company ties.[15]

Spread over five to ten years, the sums involved were relatively small in comparison to the big-money deals that were enriching some university-based entrepreneurs. And they were trivial compared to the recompense in the upper levels of the drug industry. But these and similar reports about other managers and scientists at the saintly NIH evoked indignant responses from within the scientific-medical establishment. "Damn it, if you work for NIH you're not working for a drug company, you're working for the public," Phil Lee, who served as assistant secretary of health in the Johnson and Clinton administrations, told the *Los Angeles Times*. Lee added, "When you have people who have a split allegiance, undisclosed to the public, to me it is just unthinkable."[16]

At a Senate hearing in January 2004, Katz, Germain, and other NIH officials indignantly denied wrongdoing. Earnestly professing a dedication to medical research and health care in a career that spanned nearly thirty-two years, Katz distilled the evolution of science for sale in a few words of testimony before a standing-room audience:

> In my role as physician, scientist, and leader at the NIH, I have had numerous interactions with scientists in the private and public sectors, including those in industry, and have always abided by government rules regarding such contacts. I have consulted with industry at various times beginning in 1986, when such interactions between government and industry were encouraged by then President Reagan to promote technology transfer from government to the private sector. When I became director of the NIAMS [National Institute of Arthritis and Musculoskeletal and Skin Diseases] in 1995, I conferred with NIH ethics officials and, on their advice, stopped all my consulting activities. In late 1995, I was informed that a new policy had been adopted by the NIH, initiated by then Director Harold Varmus, which again permitted such consulting arrangements. Thereafter, I began to accept consulting relationships on a limited basis.[17]

Willman duly reported that "announcing such ties is not required by the NIH," noting that "the agency has encouraged outside consulting and has allowed most of its scientists to file confidential disclosure forms." This revelation grated on scientists in universities, who felt harassed by the spreading network of rules, federal and local, designed to tame conflicts of interest. Adding to the resentment was wider awareness of selectively higher pay scales for many senior NIH employees. Their enhanced salaries were not concealed from public scrutiny, but as with so much else in the vast U.S. government, a great deal of publicly available information remains unseen because it is difficult to ferret out. To keep the NIH competitive in the biotech and medical markets, Congress authorized special salary schedules that allowed a maximum of $200,000 a year for scientists, physicians, and other high-income health professionals. The amount was above the remuneration for virtually all other federal employees but the president and vice president, and on a level with or better than the pay for many senior teaching and research positions at major institutions. Though wary of a voter backlash if it raised its own pay, Congress accepted the pleas for higher pay for research and medical specialists because of its esteem for the NIH. In a few cases, even the $200,000 ceiling was topped off by retention bonuses in the neighborhood of $50,000.

The aura of self-sacrifice in furtherance of science and the public interest sharply conflicted with the revelations of respectable pay and seemingly furtive financial dealing at the great biomedical research center. It was all either legal or arguably legal, but not easily comprehensible. The statutory basis for the higher pay was Title 42 of the Public Health Service Act, which allowed government health agencies to exceed standard federal pay levels to recruit PhD scientists for biomedical research. In use since at least the 1960s, Title 42 authority was eventually expanded to cover all health professionals, and the salary ceiling of $200,000 was established. Like other NIH employees, the beneficiaries of the higher salary scale were required to file disclosure statements detailing outside income. But, under an interpretation provided by the federal Office of Government Ethics, the forms filed by Title 42 employees were confidential, not available to prying inquirers, or subject to the Freedom of Information Act.[18] The details of outside income were on file at the NIH, or were supposed to be, but were unavailable to outsiders and of little or no interest to NIH management. In 2005, in an interview with *Science*, Zerhouni plaintively noted the ticking bomb he inherited when he became director nearly three years earlier. Addressing the 1995 relaxation of consulting rules, he said:

> Maybe [it was] justified in the general sense of saying we need to recruit and retain [good scientists]. But I think people should have realized this was a vulnerable way to do business. To not have any sense of who was doing what in a world that has changed a lot in terms of outside activities. . . . The way it was managed—don't ask, don't tell, let it be, no peer review, no disclosure of the amounts, there was nothing.[19]

The sense of things gone wrong was compounded by the finding that the profitable outside dealings of many of the figures in the embarrassing public spotlight were legitimately shielded from public disclosure. Despite its dedication to openness, in the matter of moonlighting opportunities for its scientists and administrators, the NIH utilized a selective approach to transparency. Bethesda was not immune to the tides of commercialism sweeping over the biomedical-research enterprise. The rules were rewritten to accommodate the pursuit of money.

Financing Its Competitors

Initiated in 1995 under NIH director Harold Varmus, new rules governing conflict of interest and disclosure of business dealings were adopted in response to the swiftly changing economics of biomedical research. The NIH found that it was competitive at the starting and mid-career levels of recruitment, but less so for scientific superstars and the senior scientists, physicians, and administrators who knew how to manage big, complex research organizations and programs. Given the uncertainties of research and the difficulties of managing cadres of strong-willed, independent-minded scientists, experienced administrators were especially in demand. The NIH's loss of competitiveness for these people was caused by the burgeoning academic "arms race" for scientific eminence, abetted by the rapid increase in commercial biotech opportunities and venture capital for entrepreneurial academics. Many opportunities beckoned the high achievers of science. Ironically, a major contributor to the superior competitiveness of university-based science was the growth of NIH support for research in these institutions. Across the nation, universities were rapidly expanding their laboratory facilities and research staffs to improve their competitiveness for NIH grants and scientific glory. The NIH, once a prized destination for ambitious scientists, was lapsing into competitive disadvantage. Senior staff trickled away in response to university offers of higher pay, ample lab space, and money for equipment and technicians. Drawn to the leading laboratories, topflight graduate students and postdocs, the workhorses of modern science, followed them. The wherewithal was provided by the rising NIH budget for research in universities. Thus, the NIH was boosting the competitiveness of the very institutions that were siphoning away its superior scientific talent. Harold Varmus, an alumnus of the NIH who went on to share a Nobel prize at the University of California, San Francisco, returned to the NIH as director in 1993 with marching orders to restore luster to the fading star of government science. With the revival campaign came recognition of a needlessly high level of purity in the NIH's regulation of outside activities by its staff, relative to how the rest of the government treated these matters. The disadvantage in recruiting was of the NIH's own making. As described in the report of the Alberts-Augustine Blue Ribbon panel:

> From 1988 to 1995, NIH had more stringent limits on the outside activities of its employees than it does today. In a 1995

audit of the NIH ethics program, OGE [Office of Government
Ethics] identified several restrictions on outside activities that
went beyond the restrictions in the 1993 OGE government-
wide regulations. . . . Subsequently, on November 3, 1995, the
Director of NIH notified institute and center directors and
the Office of the Director staff that NIH's outside activity
policy was being changed to conform to the less restrictive
government-wide standards of conduct.[20]

The new rules, announced by Varmus in November 1995, virtually eliminated restrictions on business dealings between senior NIH
administrators and pharmaceutical and biotech firms, and terminated
the transparency once required for such deals. With prior review for
conflicts of interest, the revision stated, NIH senior administrators

may perform the same type of outside activities as all other
NIH employees. . . . Employees may accept stock as payment
for outside activities. There is no longer a dollar limit on the
amount of income that can be received from activities performed for one or more outside activities. Employees may no
longer be limited in the amount of time they devote to activities performed for outside organizations.[21]

Outside dealings were to be examined for conflicts of interest, but as
the inspector general's inquiry later found, the NIH did not assign a
high priority to the task. As with so many other difficulties that have
emerged on the ethical boundaries of science, close attention to regulations and rules was secondary to the performance of research and
advancement of careers.

In academic and scientific folklore, sunshine is prescribed as the
best disinfectant against unsavory dealings. Under the new regulations,
dimness, and sometimes darkness, descended on the outside business
dealings of the savants entrusted to manage or conduct research at
the NIH. How did this happen? In a mind-numbing tutorial on the
changed rules of disclosure, Marilyn L. Glynn, acting director of the
Office of Government Ethics, explained at a Senate hearing that twi-
light began with the creation in 1997 of a new government job category
that permitted higher pay for recruitment and retention of sought-after
personnel, the Senior Biomedical Research Service. The new service
provided higher pay than the existing special job category for highly
valued employees, the Senior Executive Service. Employees in the new

category, she said, came under a "pay band" system, rather than the customary federal salary system of steps and grades. Descending further into obscurity, Glynn told the senators that "no employee compensated under the 'pay band' system is required to file a public disclosure report, regardless of the actual amount they are compensated. As a practical matter," she said, "this means that some employees at NIH who had been required to file a public financial disclosure report because they had previously been in the Senior Executive Service were no longer required to do so." [22]

The ground rules governing the permissible and impermissible in outside dealings by NIH staff members were a patchwork of statutory amendments and agency regulations, laxly attended to within the bureaucracy. In the absence of visible problems and outside scrutiny, the complex rules rated little attention. Now, as the NIH writhed under the *Los Angeles Times*' reports, the Blue Ribbon panel reported:

> In its interviews with NIH scientists, the Panel observed that a heightened scrutiny with regard to ethics issues has increased the confusion about the existing policies. There is a widespread sense that the rules on all outside activities are being changed midstream or suddenly overly interpreted out of caution. NIH scientists are concerned that they might not be able to fully participate in the community of science in the future, and senior management worries about the impact that possible new policies could have on the recruitment and retention of scientists at NIH. Worse yet, there seems to be widespread fear of committing an inadvertent transgression in this complex of sometimes arcane rules and interpretations. In short, many scientists sense that they are unfairly being forced to live under a cloud of suspicion. [23]

Zerhouni Stumbles

Elias A. Zerhouni was a sound choice for heading the NIH in placid times, but not in a time of unprecedented turbulence at the organization and deepening worries on Capitol Hill about the integrity and management of Congress's favorite agency. With a few hundred dollars in his possession, Zerhouni came to the United States in 1974 as a twenty-four-year-old medical graduate of the University of Algiers School of Medicine in his native Algeria. Starting with a residency in radiology at the Johns Hopkins School of Medicine, he achieved a storybook career, rising to professor and chairman of radiology and executive vice dean of the medical school. Along the way, he became a millionaire, acquiring assets in the range of $10 million to $30 million from inventions in radiology and start-up companies, according to a profile in the *New York Times*.[24] (The financial disclosure forms for government employees do not require exactitude in matters of wealth but, rather, call for stating ranges.) In May 2002 President Bush named him to fill the NIH's top position, vacant since Varmus voluntarily stepped down near the end of the Clinton administration. The NIH directorship was a sensitive post, linked to the gathering political storm over human embryonic stem cells and "right to life" and "freedom of choice" issues that Bush had introduced into the election campaign and followed up in August 2001 with a restrictive edict on stem cells. At his Senate confirmation hearing, Zerhouni avoided the stem-cell controversy, testifying that he had not been subjected to a so-called litmus test on the increasingly contentious issue. In the following year, he deftly maneuvered between the administration's ban on federal funds for research on additional lines of stem cells and the rising demands of many scientists for a larger supply. At present, he diplomatically stated, the available stem-cell lines are adequate; if more are eventually needed, he vaguely assured inquiring legislators, the need would be addressed. But when the consulting storm broke over the NIH, Zerhouni's unfamiliarity with Washington and his lack of political seasoning became evident.

The NIH has long been an anomaly in the U.S. government. Ostensibly, it is just another agency of the U.S. government, part of the cabinet-level Department of Health and Human Services. HHS, like other parts of the executive branch of government, reports to the president, who selects its leaders, sets its policies, and annually draws up its spending plans for consideration by Congress. But over the post–World War II decades, biomedical research evolved as a bipartisan favorite on Capitol Hill. Money for medical research responded to popular hopes for relief from disease, pain, and infirmity. The good-news apparatus of medical science played its part, regularly producing success stories about the latest scientific findings and expressions of need for still more money. Congress, in effect, took command of the NIH, establishing political lifelines that ran directly from Capitol Hill to the Bethesda headquarters, making the NIH an exception to the prescribed White House dominance in the budgeting and management of executive branch agencies. The NIH's congressional angels were celebrated in the biomedical community with awards for scientific statesmanship, and a succession of new buildings that Congress enthusiastically financed on the Bethesda campus were named after congressional benefactors. This was usually at congressional insistence, but with the NIH's happy compliance. Uniquely, the NIH belonged to the Congress, which exempted it from the penny-pinching theatrics that confront other agencies of government when they plead for their budgets at annual appropriations hearings. The heads of most other federal agencies are asked why they want so much and are pressed to explain how they will get along with less. NIH officials are urged to explain why they're not seeking more and how they would spend additional funds. The remarkable doubling of the NIH budget from 1998 to 2003 was initiated on Capitol Hill and successfully carried through, despite foot dragging by both the Clinton and Bush administrations. Believing that the NIH was already well-financed and hogging civilian research spending, they didn't favor it, but they couldn't resist it.

New Rules

The report of the Blue Ribbon panel, delivered to Zerhouni in draft form in May 2004, contained a series of recommendations that focused on what appeared to be the nub of the problem—unrestrained commercial activities by senior decision-making members of NIH management. Apart from these officials, "with careful review and monitoring," the panel concluded,

> it is advantageous for NIH and for the scientific enterprise to allow many NIH employees (especially intramural investigators) to engage in limited, remunerated outside activities, including those with biotechnology and pharmaceutical companies. However, the Panel recommends that other employees, specifically those in senior management positions across the institutes and centers and designated NIH extramural staff should not be allowed to engage in consulting activities with biotechnology and pharmaceutical companies under any circumstances.[25]

As new findings emerged of undisclosed, dubious dealings between NIH employees and industry, rumors circulated that punitive reductions of the NIH's budget might be in the offing on Capitol Hill. A sense of persecution enveloped the onetime nirvana of the medical sciences. Ominously warned by the NIH's long-standing supporters in Congress to clean up the mess in Bethesda and restore public confidence in his organization, Zerhouni responded to the *Los Angeles Times*' revelations with the announcement in February 2005 of draconian prohibitions against outside consulting by all NIH staff members, high and low. Ownership of more than $15,000 in pharmaceutical and biotechnology stocks by NIH staff and their family members would be banned, he said. The new restrictions extended to most honoraria and cash prizes from industry and universities holding NIH grants, as well as to other financial connections commonplace in contemporary science. Lecturing, teaching, and participation in scientific conferences would be permitted, but, as proposed by Zerhouni, the NIH would be sealed off from the commercial sector, both professionally and in major financial holdings by its employees. An uproar ensued on the normally tranquil Bethesda campus. The stock-divestiture requirement, staff scientists complained, would apply to NIH researchers who had no dealings with outside organizations, and even to technicians and other support staff. With some of the proscribed stocks trading low, divestiture would entail financial losses, especially galling for scientists and others who were not involved in consulting or in any relations with pharmaceutical firms. Scientists feared long-standing perks would be abolished in Zerhouni's blunt assault on conflicts of interest, whether real, merely perceived, or only theoretically possible. In the past, NIH scientists traveling abroad at government expense to attend scientific conferences could legitimately tack on a stretch of foreign holiday at their own expense before returning home. At no additional cost to the government, they were spared paying for plane fare. The Zerhouni edict seemed to outlaw that simple, much-appreciated gratuity of the scientific life. Several senior scientists announced they would depart the NIH for the relative serenity of academe. Others said they were looking hard. Some new recruits bound for high positions at the NIH reversed course, saying they would remain in their old jobs rather than join the oppressed NIH, where morale, by all accounts, was in tatters.

Assailed on Capitol Hill for laxity, Zerhouni now found himself accused of overreacting to the disclosures of profitable dealings by NIH employees. "Too Strict at NIH" was the title of a *Washington Post* editorial that praised Zerhouni for initially proposing a

> well-balanced set of restrictions that would have prohibited such outside arrangements for top officials and dramatically curtailed consulting by others. The final rules . . . however, go much further, imposing an absolute ban on outside consulting

and an entirely new set of prohibitions on stock ownership. . . . The stock rule, however, strikes us as too broad. It would be a hardship on numerous employees, requiring them to sell the covered stock, and probably harm recruiting as well for the same reason.[26]

The unkindest slap at the reeling NIH came from the Association of American Medical Colleges, the Washington-based lobby for the many medical schools that thrived on the NIH's money. Their researchers, perennially engaged in the NIH's competitive grants derby, looked enviously at the reliable stream of funding that the NIH provided for the scientists in its own laboratories. As the NIH moved toward strict conflict-of-interest regulations for its staff scientists, attention turned to conflict-of-interest enforcement for the academic recipients of the NIH's largess. The British journal *Nature,* which routinely counsels scientists in the former colonies on proper behavior, editorialized that "scientists and institutions everywhere should be sure that their own houses are fully in order," noting that Zerhouni himself had indicated in a recent speech that his concern with conflict of interest extended beyond the Bethesda campus.[27] The clear implication evoked a chop-logic rejoinder from AAMC president Jordan Cohen, endorsing the ban on consulting by NIH employees, but opposing its extension to scientists in universities.

The AAMC president, who retired in 2006, addressed the reports of financial improprieties by government employees from his own financially well-padded, nonprofit perch. As AAMC president, Cohen, a former medical school dean, received total compensation from the association of $896,836 in 2003, according to the publicly available, nonprofit tax return filed by the AAMC. In 1996, the tax return states, the AAMC provided him with an "Interest Free Loan" of $765,965 for purchase of a residence, with "Imputed Interest Charged as Income to the Borrower."[28] For NIH employees, pay at that level—more than double the salary of the president of the United States—plus mortgage assistance, was unavailable, unthinkable. Writing in his regular column in the AAMC house organ, Cohen observed that

not surprisingly, some in the media and elsewhere are now asking why these extremely stringent rules should be limited to the intramural scientists at NIH? If it's appropriate for them to be heavily insulated from financial temptation, they ask, why not apply the same rules to university scientists? The answer is that university scientists operate in a much different environment than their NIH counterparts, and play a much broader societal role. . . . The principal difference centers on the public's interest in fostering productive partnerships between academia and industry. Beginning long before the enactment of the Bayh-Dole Act of 1980 . . . such partnerships have been widely recognized as essential, not only to the vitality of university science and science education, but also . . . to accelerate the translation of new knowledge into useful products and services. To that end lawmakers have put strong financial incentives in place to encourage academic institutions and their faculty to team up with for-profit entities in order to harness their unique capacity to fully develop promising ideas into actual benefits. . . . Accordingly, both the federal government and the academic research community have determined that the public's interest in gaining rapid access to the benefits of medical research would be best served by managing, rather than banning, all financial interactions between academia and industry. Research universities have had long experience at managing faculty conflicts of interest across the broad range of scholarly disciplines. Conflicts in biomedical research have received special attention, particularly since the issuance of mandatory federal regulations for the extramural community in 1995 and their enhancement by comprehensive federal guidelines in 2004. . . . Indeed, the critical lesson to learn from the NIH experience is that strict compliance with appropriate conflict of interest standards is the sine qua non for preserving public trust in the integrity of our research.[29]

Cohen's argument ignored the fact that in 1980, the year that Congress passed the Bayh-Dole Act to promote technology transfer from universities to industry, it also passed the Stevenson-Wydler Act (PL 96-480), to promote technology transfer from government laboratories to industry. In 1986 a companion law, the Federal Technology Transfer Act (PL 99-502), authorized royalty sharing for government scientists whose inventions were successfully marketed by industry. For moving government-financed scientific results to industry, federal law does not distinguish between government and academic scientists or their organizations.

August 2005 brought the final version of the new rules for consulting and financial holdings by NIH staff members. Arrived at after the receipt of some thirteen hundred comments, bitter protests from many NIH employees, resignation threats, recruitment difficulties, and several departures from the NIH, the regulations were strict but considerably less so than those proposed by Zerhouni six months earlier. Moreover, their duration was stated to be one year, which meant they could be continued as is, modified, or dropped. Under the new rules, consulting for biotechnology and pharmaceutical firms was banned for NIH employees at all levels. But only senior employees and their immediate family members would be prohibited from owning more than $15,000 worth of stock in any such firms. Lower-level employees would not be automatically barred from stock ownership, but many would be required to disclose their holdings and, if deemed necessary following a review process, might be required to sell specified stocks. Many of the earlier restrictions on lecturing and other outside activities were modified or eliminated. However, Zerhouni's and the NIH's deep suspicion of the commercial interests hovering over biomedical research was evident in an NIH explanatory document accompanying the new regulations. Regarding writing and editing for scientific journals and textbooks, the explanation stated: "This kind of activity will not be permissible if a pharmaceutical or biotechnology company funds the publishing activity unless it does so as an unrestricted financial contributor and exercises no editorial control."[30] For participation by NIH staff in continuing medical education programs, similar hands-off rules applied to company funding.

The new rules, issued in final form in August 2005, relaxed some of the restrictions on stock holdings, but prohibited virtually all of the income-producing consulting activities that had brought shame on the NIH. However, the regulatory laxity and confusion that had abetted the profitable links between NIH scientists and pharmaceutical firms was undeniable, and for most who were snared in these dealings, the disciplinary consequences were slight to nonexistent. A major exception was P. Trey Sunderland III, the NIH mental-health specialist who pleaded guilty in federal court in December 2006 to violating federal conflict-of-interest regulations. Sunderland admitted that he had failed to obtain authorization for $285,000 in consulting fees and $15,000 in expenses from Pfizer, and entered into a plea agreement to pay $300,000 to the government and perform four hundred hours of community service.[31]

Six other scientists quit the NIH before any action was taken against them, leaving uncertainty about how they would have fared had they stayed on. Twenty-eight others received letters of "caution," "admonishment," or "reproval." Five received short suspensions with pay; one a forty-five-day suspension without pay. At a congressional hearing, NIH officials said Sunderland and one other scientist who had received large consulting fees held commissions in the Public Health Service and remained employed at the NIH because they were beyond the NIH's disciplinary reach.[32]

The NIH came out battered and demoralized from its entanglements with commercialism, which was never widely popular among the scientists who opted for careers in government service. Zerhouni's absolute ban on consulting and restrictions on stock ownership reflects a fear that honest science risks contamination from commercial contacts, that the danger is so great that prohibition rather than management is the only feasible solution. There was no resort to diplomatic language. In direct contrast to this stark view of industry as a menace to scientific integrity, the AAMC president Cohen contended that universities can collaborate with industry, gain riches, serve the public, and retain their virtue. Zerhouni's doubts are on the record. In his interview with *Science,* he critically addressed some of the money-making practices that continued to be tolerated in academic science, even in the new era of ethical sensitivity. Referring to the practice of medical-school mercenaries shilling for pharmaceutical manufacturers and other disreputable but common moneymaking tactics of university scientists, he said:

> Preserving science is, I think, a discussion that we need to have. Is it okay to be on the payroll of the marketing department [of a drug company] and to make thousands of dollars going to medical meetings and saying, "I'm a scientist, I'm very trusted, I'll tell you what, this drug is better than this drug"? The trading of scientific credibility units for dollars for a marketing or promotion goal is something that we need to talk about. . . . We are seeing a very worrisome trend in the trust factor of science, especially when it comes to human subjects science.[33]

Human experimentation has long been a troublesome sector of scientific research, especially when money is riding on the outcome.

6   Conflicts and Interests

Because of the way the system was set up, it was a system where the government said, "We'll give you the money and you have to promise that you'll
meet these obligations for protection of human subjects, management of
funds, conflict of interest, research integrity, whatever." So that this assurance process that was put in place was largely a paper process that for the
first twenty years of OPRR's [U.S. Office for Protection from Research Risks]
existence rarely went beyond the exchange of paper back and forth unless
something bad happened. And then someone would look and say, "Well,
you screwed up." The critical change that occurred was when OPRR actually
started going out proactively to do these site visits. So what they discovered, and it's a perfectly understandable phenomenon, when the government
says, "Okay, we'll give you the money, you promise to do it, we're not going
to look and see," it provides an incentive for institutions to do as little as
they can to meet the letter of the law. And, in fact, if nobody's looking, why
even do that?

Greg Koski, associate professor, Harvard Medical School; director,
U.S. Office for Protection from Research Risks, 2000–2002[1]

: : :

The protection of people who serve as subjects of medical
experiments is a fundamental requirement of scrupulous
scientific behavior. This principle, enshrined in contemporary science, leads to an unrelenting focus on conflict
of interest among the administrators and researchers
running clinical trials. The use of humans in experiments is necessary for satisfying the FDA's requirement

that drugs and medical devices are safe and effective. Humans are also required for research directed at understanding basic biological processes, a frequent preliminary to product development and testing. By law, participation in experiments must be based on "informed consent" and the expected risks must be "reasonable in relation to the anticipated benefits."[2] Furthermore, under federal regulations in effect since 1995, government-financed researchers in universities and other nonprofit organizations must disclose to their institutions any "significant" financial holdings—usually defined as $10,000 or more—that might be affected by their projects. Many difficulties, evasions, and misdeeds have arisen from these requirements, along with bitter allegations that the pursuit of money often undermines ethical science.

The Nuremberg Code

After World War II, the discovery of gruesome Nazi experiments on helpless humans established a base of revulsion at science gone amok. From the trial of Nazi physicians came the Nuremberg Code, a statement of medical ethics. Written by American physicians assisting the prosecution, the ten-part code was incorporated into the verdict under the heading "Permissible Medical Experiments." The opening section mandates "voluntary informed consent." Spiritually, though not explicitly, incorporated into American law and medical practice, the Nuremberg Code became a landmark in the evolution of medical ethics and served as the starting point for further declarations and codes, national and international, in behalf of humane science. In 1991 a single set of provisions governing protection of human subject—the Common Rule—was adopted by sixteen U.S. government departments and agencies.

The Nazi medical crimes were faraway horrors, committed under a brutal regime, bearing no relation to the humanitarian aura surrounding medical science American style. However, the realization that it does happen here arrived when misdeeds by American scientists and physicians became publicly known. Though not remotely on the scale of the Nazi crimes, they were nonetheless shocking. In 1964 widespread, indignant media attention was given to reports that elderly, debilitated patients in a New York hospital were experimentally injected with cancer cells without their consent or knowledge. Later came reports of experiments on prison inmates in dubious ethical circumstances and other cases of human "guinea pigs" who were lured or unwittingly drawn into medical experiments.

Politically and psychologically, the decisive revelations came in 1972 with exposure of the infamous Tuskegee experiment. This involved 412 black men who had been diagnosed with syphilis and 204 uninfected men, so-called controls, enrolled in a U.S. Public Health Service research project that began in 1932. The participants, mostly poor sharecroppers, were recruited for the PHS by the Tuskegee Institute, a black university that commanded respect and attention among blacks in the surrounding Alabama area. After the standard drugs of the time—arsenic, bismuth, and mercury—proved ineffective, treatment was halted. Penicillin, effective against syphilis, became available in the mid-1940s but was withheld from the infected men so that researchers could study the natural course of the disease. In the furor that followed public disclosure of the project, the U.S. government agreed to a $10 million out-of-court settlement of a class-action suit. The mad scientist, à la Dr. Frankenstein, has long lurked in the nightmares of technological society. Now, as reality matched fiction, the luster dimmed on the beneficent halo of science.

It would be incorrect, however, to conclude that these dismal episodes seriously diminished public or political confidence in science. In 1972, the same year that the Tuskegee affair became public knowledge, Congress began generously funding the newly declared "war on cancer," which commenced with President Nixon's signing of the National Cancer Act in December 1971. Promoted with rash assurances that blank-check spending would eventually bring victory against the disease, the new law demonstrated society's increasing ambivalence toward science: hope was tinged with concern, even fear.

In 1974 Congress passed the National Research Act (PL 93-348), which led to the establishment of institutional review boards (IRBs) at universities and other research centers to ensure the protection of people in medical research projects. The law required that before a government-financed research project involving humans could proceed, it would have to be scrutinized by the boards for adherence to high ethical standards. Documentation establishing the voluntary, informed participation of research subjects was crucial, as were data derived from animal studies and prior publications relevant to minimizing the risks. Recognition of the potential, if not the likelihood, for poor compliance or evasion within research institutions was evident in the membership criteria for the boards, which specified that they were not to be wholly comprised of an institution's own employees. Outsiders were to be involved, too. Even so, the rules were far from strict and enforcement was faint to nonexistent, reflecting political reluctance to intrude on higher education or science. Washington was by no means putting tough cops on the ethical beat in the nation's research universities.

Of the required minimum of five members for each board, at least one was to be drawn from outside of scientific research and one was to be unaffiliated with the institution where the research was to take place. The others could be employees of the institution, but any vote required the presence of a nonscientist member. The obligatory inclusion of a nonscientist and an outsider suggested a lack of confidence in the self-governance of science, but within the stated criteria, the university was free to fill these roles as it pleased. Nonscientist members of the university and supportive outsiders were eligible. Animal-rights activists and other carping critics of science were also eligible, but not likely to be picked. The rules did not provide for external oversight or verification of the boards' diligence or consideration of conflicts of interest involving the board members and the research they would review. Nor did the rules require the boards to consider the financial interests of the institutions where the research was performed or of the researchers who recruited the volunteers, conducted the experiments, and reported the results. The job of the IRBs was to certify that people serving as experimental subjects knew what they were doing when they signed on and that the research would not expose them to unreasonable risks relative to the expected benefits. The flimsiness of the federal regulations was acknowledged in 2001 by a committee of the Association of American Medical Colleges, an organization instinctively averse to government intrusions on scientific independence. The rules, it said, do not require disclosure of financial involvements by university researchers experimenting on humans, "nor do they acknowledge the unique obligations that attend research involving human beings."[3]

A Collegial System of Safeguards

Researchers with a financial stake in the outcome of their experiments were a growing presence in universities and medical centers. Moreover, the institutions themselves sometimes held a financial interest in clinical research on campus through the venture capital they provided for professorial start-up companies, through pharmaceutical stocks in their endowments, and other holdings and business dealings. If they wished to, the IRBs could consider financial conflicts of interest in their deliberations, but few chose to, and at most institutions the issue was either ignored or handled by a separate conflict-of-interest committee. From the start of the IRB process in the mid-1970s, until the 2003 blowup over consulting by NIH employees, the prevailing sentiment in the governance of research held that financial conflicts of interest are ubiquitous in modern professional life, generally harmless to scientific integrity, and tolerable or manageable. Though far outweighed by federal funds, commercial money in the academic research system was so familiar that in 2000 at an NIH conference on conflict of interest, a speaker scoffed at ethical concerns, wisecracking that "$25,000 these days won't buy a lot of bias anyway." [4]

As reflected in longtime NIH practice and attitudes, and eventually specified in NIH regulations for grantees, disclosure of financial interests to university authorities was considered a sufficient guarantee of good behavior in most instances. In some circumstances, "monitoring of research by independent reviewers" might be necessary. In situations judged to be deeply contaminated by financial interests, removal of the conflicted scientist from the project or divestiture of problematic financial interests was the recommended, but rarely used, antidote. [5] The system established by the National Research Act was collegial, not adversarial.

Typical of the anti-government streak in American society and politics, oversight of the boards and enforcement of the regulations were essentially left to the institutions where the research was conducted. Washington tended to gentleness in its relations with universities, especially the major research institutions that received most of the NIH's grants. Entrepreneurial values and minimization of government regulations received strong backing. In 1980 the U.S. Supreme Court, in *Diamond v. Chakrabarty* (447 U.S. 303), ruled five to four that human-made microorganisms were patentable—a revolutionary, and highly controversial, decision that opened a new frontier in scientific-commercial dealings. In 1981 Bayh-Dole took effect and, coincidentally, marvelous discoveries in biotechnology, with enticing clinical potential, were coming out of university laboratories. And that was the year in which President Ronald Reagan told his first inaugural audience that "government is not the solution to our problem; government is the problem. . . . It is time to check and reverse the growth of government. . . . It is my intention to curb the size and influence of the federal establishment." Federal employees responsible for enforcing the IRB rules understood the message. With heavy volumes of government money flowing into the research system, scientists were eager to get on with their work and resentful of government-imposed bureaucratic hurdles, such as IRBs.

The new rules required clinicians to report to the NIH or the FDA "adverse events"—that is, patients sickened or, as rarely happened, killed, or experiments otherwise gone awry in the course of testing drugs or other procedures involving humans. Researchers were thus obligated to confess to failures that might otherwise remain unknown to faraway federal authorities. Amid the competition for research money and career advancement in science, this invitation to step forward and blot one's own reputation did not engender strict compliance. Not uncommonly, failed experiments would be quietly terminated, and the data collected along the way, though possibly of scientific or medical interest, deposited down the memory hole. The same fate frequently awaited so-called negative results from drug tests conducted for pharmaceutical firms. Though these findings, too, could be useful, drug companies preferred not to acknowledge failed trials of their prospective drugs. Academic researchers knew that the scientific and medical journals essential for advancing their careers had little interest in publishing reports of experiments that failed to confirm the underlying hypotheses. Thus, reports of action on the endless frontier were skewed in favor of success, while failed outcomes usually went unrecorded. That a lot of research effort produces nothing of recognizable value is one of the less-advertised facts of the research enterprise. Through this filtering process, the professional publications teemed with success stories. Since popular science writing feasts on medical and scientific journals for basic material, the public and politics were led to believe that continuous, rapid progress came from big spending on science, thus supporting appeals for even bigger spending. Still there was never enough.

Pressed for money, as always, universities resented Washington's refusal to pay directly for the costs of running the IRBs, such as clerical help for the extensive record keeping required to do it right. Computerization of the records advanced slowly because of the large volume of paper generated by research projects, especially when humans subjects are involved. At several universities I saw thousands of feet of shelf space occupied by thousands upon thousands of folders stuffed with documentation required by the IRBs. These included copies of grant applications, commonly filling scores of pages, correspondence from government program officers seeking further information and correspondence providing it, prior publications relevant to the proposed research, explanatory material for prospective experimental volunteers, informed consent forms, the IRB's initial and periodic assessments of projects, and much more paper, all collected and stored on the assumption that it is better to store it and not need it than to need it and not have it.

Some financial help for IRB operations could be derived from the indirect-cost payments that accompanied government research money, but federal bookkeepers kept a close watch on these university expenditures, and eventually a tight cap was put on the amounts that could be attributed to administrative services. IRBs were a money loser in an academic economy increasingly sensitive about profit and loss centers. Universities' repeated pleas for federal financing of IRB operations failed to impress Washington, where the denial and provision of money is often inconsistent and mysterious. The sums involved were relatively small in the context of overall budgets at academic medical centers, and they certainly were negligible in relation to widespread concerns about patient protection. According to a 2002 survey of IRB costs, the median for all schools was $741,000 each per year. For schools with low volumes of research, the median was $402,369; for high-volume institutions, $1,150,417. Staff salaries accounted for most of the costs.[6] Though scarcely onerous, these expenditures offended academe's budget managers. The constant grousing was another manifestation of the enduringly straitened condition of academic finance.

The time required for performing IRB duties was another irritant. Service on the boards was voluntary and unpaid, an act of good academic citizenship, but yet another committee burden in the academic schedule. Moreover, service on an IRB risked run-ins with colleagues whose research proposals were found lacking in patient protection or some other aspect of the stricter emphasis on research ethics. The conflicts and frictions inherent in the IRB system were drolly described in 2000 by a longtime observer, Jeremy Sugarman, then of Duke University:

> An institutional review board has to approve research. If it doesn't no grants, no contracts come in. With no grants and with no contracts, there are no direct costs. With no direct costs or no indirect costs—if there are no indirect costs, there are no doughnuts for the IRB meeting and everything closes down. This is an inherent conflict of interest that was recognized from the very beginning. . . . There are problems with having to work with superiors and inferiors and colleagues and friends and enemies. The stakes are always so small in academia . . . the competitions are so strong. We have recognized this conflict from the beginning.[7]

Money is the infallible measure of what government considers important. The federal Office for Protection from Research Risks

(OPRR), responsible for monitoring human and animal research financed by the NIH, was budgeted for $2.6 million in 1999, while the NIH was running on $15.6 *billion* and supporting some thirty thousand research projects at thousands of locations. Situated at subterranean depth in the NIH bureaucracy, OPRR's staff of twenty-seven included only one full-time investigator, one full-time attorney, and one physician, serving part-time. Eight members of the staff monitored the treatment of experimental animals. Most of the time, most of the staff were occupied with paperwork at the home office, near Washington, D.C. The director, Gary Ellis, told me that from 1991 to 1999, OPRR performed only thirty-eight inspection visits to universities, and during one thirteen-month period, no inspections.[8] The NIH reaped scientific renown, huge congressional appropriations, and popular support by financing research in universities and in its own laboratories, in a noble and well-publicized war on disease. Snooping on the ethical performance of its warriors held low status in the NIH culture and in the distribution of its soaring but never sufficient budget. To make its way in Washington, a regulatory agency needs a supportive constituency. The Federal Aviation Administration has the flying public. The Environmental Protection Agency benefits from nationwide "green" sentiment. The tiny OPRR was backed by a few concerned academic scientists, who hoped it would serve as a guardian against ethically lax professors and the temptations attached to commercial money surging into clinical research. It also drew support from several patients-rights groups founded by relatives of volunteers who had suffered or died in experiments. Otherwise it was friendless, virtually unknown in Washington, and, at best, only grudgingly accepted by the majority of academic researchers and their university administrators.

For the watchdogs of biomedical ethics, Tuskegee provided a fundamentally important message: the conduct of human experimentation could not be entrusted to the good intentions of scientists. But elsewhere in society, memories of Tuskegee generally receded.* The

---

*The relatively low participation of African Americans in clinical research studies has been attributed to a residue of distrust arising from the Tuskegee episode. But a large-scale study disputes that interpretation and attributes the differential in enrollment rates to insufficient outreach to minority patients, lack of child care, and poor transportation facilities in minority neighborhoods. "We found very small differences in the willingness of minorities . . . to participate in health research compared to non-Hispanic whites. . . . Hence, efforts to increase minority participation in health research should focus on ensuring access to health research for all groups, rather than changing minority attitudes." David Wendler et al., "Are Racial and Ethnic Minorities Less Willing to Participate in Health Research?" *Public Library of Medicine*, vol. 3, no. 2, February 2006.

good-news collaborative of science and the media twinkled with reports of wondrous cures, real and imagined, from the rapid expansion of the biomedical-research enterprise. Tuskegee was long ago. Attention to the intricacies of new federal rules for research was not a high priority. Moreover, when taken seriously by researchers and university administrators, voluntary informed consent, balancing of risks and benefits, and avoidance of conflict of interest became a legalistic and psychological swamp. Exploration of rights and wrongs in human experimentation nourished an academic subdiscipline, bioethics, which developed research programs and a thriving conference circuit, underwritten by public and private research grants. Government commissions and nongovernment bodies issued reports. In symbiotic harmony, the participants produced immense quantities of literature and commentaries on the literature, reflecting a seemingly insatiable fixation on their topic. In August 2006 a Google search dredged up 23.2 million entries for "conflict of interest" in combination with "research," and 21.7 million for "informed consent." Even allowing for cyber retrieval's indiscriminate grasp, the astronomical, and continuing, yields indicate that closure on the interplay of science, ethics, and commerce remains elusive.

In laboratories and clinics, the rules left many researchers puzzled and frustrated. The daunting scientific and technical difficulties of making progress in biomedical research were now compounded by ethical and legal prescriptions and ambiguities. To take a relatively simple example: Many cancer patients and others with life-threatening diseases are invited to participate in clinical trials aimed at developing or evaluating treatments. Given the stress caused by serious illness, and the generally poor scientific and medical understanding of the American populace, do these patients comprehend that in a clinical trial they may unknowingly receive the sugar-pill placebo or an older drug rather than the new "miracle" drug that they pray will save their life? Or that the new treatment being tested may prove to be inferior to an existing treatment or possibly even harmful to patients? And what are they to conclude from learning that the compassionate physician-scientist presiding over their treatment has a financial stake in its success through stock options in the company that produced the drug? Does the prospect of a bonanza from experimental success affect the researcher's judgment of the risk-benefit balance? Is it relevant that the company is desperate for a favorable nod from the FDA before the "burn rate" exhausts its start-up venture capital? A wisp of good news can send a start-up's stock into orbit and bring in another stash of venture capital.

Bad news can set off a price-busting stampede to sell. Financial inducements for signing up as experimental subjects are permissible and often are necessary to attract healthy volunteers, but at what point does money begin to undermine the "voluntary" criterion for human experimentation?

A recent survey found, not surprisingly, that cancer patients enrolled in drug trials have scarcely any interest in these financial-ethical conundrums. Based on interviews in 2004–5 with 253 patients at five medical centers across the United States, the survey team reported, "More than 90 percent of patients expressed little or no worry about financial ties that researchers or institutions might have with drug companies. . . . Most patients . . . would still have enrolled in the trial if they had known about such financial ties."[9]

With federal regulators in a seemingly passive state, and patients intently focused on treatment, biomedical researchers and their administrators comfortably opted for the easiest route across the difficult, but unsupervised, regulatory terrain. They devoted little attention and resources to Washington's IRB requirements—except for an ironical consequence of the emphasis on voluntary informed consent. In attempts by universities to protect themselves against all the litigious potentialities of experimentation, the informed-consent forms grew to prodigious length, so that anxiety-ridden patients might be confronted by and required to sign a sheaf of dense medico-legalistic text before being admitted to the treatment facility. The passage of years worsened rather than simplified the consent process. Sharon K. Friend—director of the Human Research Protection Program at the University of California, San Francisco—recalled that in the mid-1980s, the consent forms consisted of a few pages. Noting their inexorable growth over twenty years, she told me:

> The consent forms are totally out of control. By the time you get
> what all the lawyers want in them and what the [IRB] members
> want in them, the consent form for a children's oncology group
> study—these kids are dying of cancer—are averaging twenty
> pages of graduate-school-level language. It's a lot of stuff, and
> then they meet with the kids, the family members. They have
> a room with a big bulletin board and the doctors and nurses
> come in and they go through the regimen and what's going to
> happen. It's really complicated. And they're terribly, terribly
> stressful situations. And then they give them this twenty-page
> consent form. They get more than one consent form. If they

> come here, they're going to be enrolled in a study, and then they
> have a tissue bank they might want to ask them to be in, and an-
> other registry. So they can have three or four consent forms.[10]

Diligence, resources, and quality naturally varied among the three
thousand or more IRBs that were initially established to comply with
the National Research Act. But sloppy, inadequate performance of
many IRBs became a publicly known aspect of biomedical research,
beginning with the start-up period in the mid-1970s and continuing
into the twenty-first century. During those years, serious deficiencies
were repeatedly identified and publicly reported and criticized. But
when IRB performance was looked at again several years later, and
then again still later, the same failings were noted, along with new IRB
problems fostered by the burgeoning of the biotechnology industry and
an accompanying ravenous need for clinical trials. In 1982 a joint task
force of the National Cancer Institute and the Food and Drug Administration, established to investigate the protection of human subjects in
cancer research, reported that "problems continue to plague the IRBs
and their performance." Referring to an FDA review of the M. D. An-
derson Hospital and Tumor Institute, a leading center for cancer research in Houston, Texas, the task force reported:

> That evaluation revealed some commonly encountered prob-
> lems. The M. D. Anderson IRB procedures for annual review
> of clinical investigations were found inadequate in that review
> of the project level did not ensure that each study had been
> reviewed. The IRB was not informed when a study had been
> terminated, and most importantly, information on ADRs
> [adverse drug reactions] and patient population and protocol
> changes were never brought to the IRB's attention. . . .
>
> Part of this is directly related to the absence of specific di-
> rection and education of members.[11]

Thirteen years later, in 1996, the General Accounting Office, which
conducts investigations for the Congress, studied a sample of IRBs and
concluded that many were rubber-stamp operations that provided little
assurance of protection of human subjects. "In some cases," the GAO
found, "the sheer number of studies necessitates that IRBs spend only
one or two minutes of review per study."[12] Despite similar reports at
the time, a delusional sense of goodness pervaded the leadership of the
biomedical-research community. Writing in his organization's house

organ in 1998, the head of the Association of American Medical Colleges, Jordan Cohen, extolled the performance of the IRBs:

> Guided by the principles of beneficence, justice, and respect for persons, IRBs weigh the risks posed by research against the benefits that the research may offer to the patient and society. They work *collaboratively* [italics in original] with investigators, the vast majority of whom are motivated by altruism.[13]

New Issues and Mounting Criticism

Reports slamming IRB performance kept coming. In 1999 the GAO told Congress that while old problems persisted at many IRBs, new ones had developed. Among them was the willingness of drug firms to pay fees to the university-based IRBs that reviewed the ethical quality of their human-research projects while the NIH and other government research agencies persisted in refusing payment for IRB review of their clinical trials. Given academe's usual financial stringencies, how did that disparity affect the behavior of faculty researchers and administrators? Courtship of industry and an eagerness to accommodate its wishes for speedy review and approval of research plans were the outcomes, according to the GAO.

Another new factor was the spread of commercial firms that performed clinical research, so-called contract research organizations (CROs). In 1991, 80 percent of industry's money for clinical research went to universities; by 1998, the academic share had dropped to 40 percent. A report in the *New England Journal of Medicine,* noting this shift, observed, "Evidence suggests that the commercial sector completes trials more rapidly and more cheaply than academic medical centers."[14] The rise of private, non-university organizations for conducting clinical trials was accompanied by another shift from the academic to the private sector. Privately owned, profit-seeking IRBs were created to review patient safety, informed consent, and other requirements for the conduct of clinical research. They mainly worked for the private sector but also for universities that found it difficult to cope with the demanding requirements of proper IRB operations. As far as federal regulators were concerned, private profit-seeking IRBs were on a par with academic nonprofit IRBs. Both were required to abide by the same rules, but as usual, no one was diligently searching to ensure that they did. Pointing to "the commercialization of research," the GAO stated:

Industry is now the largest sponsor of clinical research in the country. With the increasing prominence of industry comes the pressure for "more, faster, better" IRB reviews. Investigators are expected to recruit subjects quickly and IRBs are prodded for timely approvals of research protocols. IRBs' own institutions are often focused on bringing in important research dollars from industry sponsors. Many see thorough IRB review as a hurdle for their research efforts. Hand in hand with the commercialization of research is the potential for conflicts of interest. It is important to note that these conflicts are prominent in all research settings—academia, industry, and independent—and from all sources of funding.[15]

The GAO warned that with the rapid expansion of research sponsored by government and industry, the creaky IRB system was unable to maintain adequate surveillance over the increasing use of cash incentives to facilitate enrollment of volunteers. Noting these developments, the GAO stated:

The intensified quest for [human] subjects heightens recruitment pressures and leaves IRBs with many difficult questions to confront. Should they be concerned about recruitment bonuses that sponsors give to investigators? What about when investigators mine patient databases to find potential subjects? Should investigators offer fees to physicians when the physicians refer their patients as subjects? Few guidelines exist to help answer such questions.[16]

Great organizational changes were occurring in the conduct of clinical research as pharmaceutical and biotechnology firms, confronted by the rising costs and regulatory complexities of academic medical centers, shifted clinical-testing contracts to non-academic physicians, private medical organizations, and, increasingly, to sites in less-developed countries—all acceptable to the FDA if the sponsors and performers adhered to the clinical testing rules. However, if these widely dispersed clinical trials cut corners to hold down costs and speed the delivery of results, the likelihood of detection was far less than in the American academic setting, with its norms of openness, ethical sensitivities, fears of shame, and the ever-present possibility of a whistle-blower causing trouble. Strict observance of IRB regulations raised the costs and administrative hurdles for universities, rendering them less competitive for conducting clinical trials, and thereby reducing the revenues that provided salaries for many of their physician staff members. The shift to foreign sites became noticeable in the early 1990s and continued to accelerate in the new century. In 2005 Thomson CenterWatch, which monitors the clinical-trial industry, forecast that

> by 2008, up to 30% of global clinical trial activity will take place outside of the U.S. and Western Europe due to high demand for study subjects and well-trained clinical research professionals. China, Eastern Europe and Latin America are several key markets earmarked for rapid growth in clinical research grants. Some estimates are that India could capture up to 8–10% of the global clinical trial market by the end of the decade. . . . Central and Eastern European countries have become increasingly popular locations for clinical research, and the interest continues to push eastward as Poland, Hungary and the Czech Republic approach the saturation point. While lower costs remain a draw, access to motivated investigators and patients concentrated in large public hospitals are the biggest advantages. Major sponsors and CROs are all well represented in the region, aided by small but capable local CROs. Trial data overall are exceptional. . . . Investigator fees, shorter recruitment timelines, lower hospital fees, cleaner data all contribute to cost saving for biopharmaceuticals doing business in CEE [Central and Eastern Europe].[17]

The long-standing as well as the new and worsening problems in surveillance of clinical research were well known within the academic research community and in government health-policy circles. But there was little motivation, incentive, or leadership to deal with them. Government regulation in this area was halfhearted and poorly financed. In the hurly-burly of the competitive research system, the fine details of ethical-research requirements rarely rated a high priority among harried scientists and administrators. They earnestly regarded themselves as good people, decent, personally and professionally committed to alleviating disease and pursuing other publicly desirable goals, at wage levels considerably below those of their peers in the private sector. Besides, everyone knew that major universities have rules, regulations, and safeguards: "fire walls," to shield purity from avarice, and "bright red lines," beyond which faculty entrepreneurs and their commercial collaborators supposedly may not go. At virtually every major university, professors were annually required to report their outside financial dealings and holdings. Some universities adopted financial disclosure requirements on their own, while others responded to the 1995 federal diktat requiring NIH and NSF grantees to inform their universities of any income above $10,000 from a company that might benefit from their research.

The requirement represented a naive faith in the power of disclosure. Though it may be inferred that outside income above $10,000 in such circumstances was frowned upon or at least looked at with suspicious interest, it was rarely forbidden, nor was the disclosure to be publicly available; it merely had to be delivered to authorities at the university. Harvard was an exception. Richer and more pious than all others, it had previously set income ceilings for its faculty of $10,000 a year in consulting fees and $20,000 in stock from a company with an interest in a faculty member's research. With many professors chafing at blocked opportunities for income, Harvard Medical School came close to raising the limits in 2000, less than a year after the gene-therapy debacle at Penn. There was nothing to prevent Harvard from setting its own limits sky-high for outside financial dealings by faculty, which could confidently expect a rush of commercial enterprises to come calling for their skills and the prestigious Harvard brand. But the potential for embarrassing misdeeds in an ethically loosened environment proved decisive.

Seared into institutional memory was a highly publicized episode in the mid-1980s involving a researcher, Dr. Scheffer Tseng, at the Harvard-affiliated Massachusetts Eye and Ear Infirmary. Tseng, members of his family, and several colleagues greatly profited from stock ownership in a start-up company that produced an eye-treatment drug for which Tseng conducted clinical trials. Investigators reported serious violations of IRB requirements and conflict-of-interest regulations. The Tseng affair was traumatic for Harvard. Shortly before a meeting with faculty was to be held to discuss a revision of the conflict-of-interest and financial regulations, Joseph Martin, dean of the medical school, called it off, explaining: "I believe that the most important role academic medicine can have in clinical research today is to try to bolster the public's faith in the veracity and ethical underpinnings of this noble endeavor." [18]

Other counterreactions to the onward march of commercialization were occurring at crucial points in the research enterprise. Slowly and long overdue, major scientific and medical journals exercised their editorial power against financial contamination of research. As a condition of publication in their prestigious pages, the journals required prospective authors to disclose any financial connections to the subject matter of their papers, and notes on such reported ties are usually published as an appendage to the article. The *New England Journal of Medicine* and *JAMA* publish the names of organizations that have provided research support and honoraria for its authors, and firms in which they hold stock. The *Lancet* takes openness a step further, requiring a "conflict of interest statement" that includes financial connections and an assurance that "the funding source had no role in study design, data collection, data analysis, data interpretation, or writing of the report. The corresponding author [the contact for readers] had full access to all the data in the study and had final responsibility for the decision to submit for publication." The statement, published when relevant, is intended to block common machinations of the pharmaceutical industry, such as experimental designs that favor a desired outcome, omission, or distortion of unfavorable findings, and "ghost" authorship.

But as with so many other regulations designed to uphold scientific integrity, awareness among scientists was spotty, enforcement was lax, and noncompliance no rarity. There's no joy in filling out forms listing all the possibly conflicted outside income collected over many years by a successful researcher. Journal editors stressed the importance of disclosure of conflicts of interest in the articles they published, but insisted they lacked the resources to track the financial dealings of their authors. The energy and ingenuity that were routinely devoted to pursuing grants and conducting research did not extend to the bureaucratic minutiae of the modern, fiercely competitive scientific life. Yes, disclosure, transparency, ethics were all very important, but . . .

Who Guards the Guardians?

Universities vary in their ardor for ethical policing. Employment of graduate students in faculty-owned start-up companies is generally forbidden or restricted at many universities out of concern about diversions from serious academic studies and commercial limitations on their freedom to publish their dissertations. A blatant violation of these principles produced image-staining embarrassment for MIT in 1999, as reported in the *Wall Street Journal* under the headline "MIT Students, Lured to New Tech Firms, Get Caught in a Bind: They Work for Professors Who May Also Oversee Their Academic Careers."[19] The article told of an MIT undergraduate who was unable to carry out a homework assignment because he was bound by a nondisclosure agreement at a company where he worked. The company was owned by an MIT professor. The student's instructor owned a competing company, which led to charges that the instructor was using a homework assignment for industrial espionage. Solution: a different assignment for the student.

Six years later, when I referred to that incident in conversations with MIT administrators, they characterized the episode as an aberration that had occurred long ago and assured me that a repetition was out of the question. Is that really so? I cannot say for certain. Universities are usually tight-lipped about disciplinary matters, often citing privacy regulations—appropriately or not. But by the late 1990s, acute sensitivity to public embarrassment was increasingly evident in the governance of academic institutions, with their heavy dependence on public approval for raising funds that are always sorely needed.

In the operations of many universities, entrepreneurial or not, anything that looked even a bit unsavory or that might blight the school's good name was likely to be brought before a conflict-of-interest committee, yet another addition to the committee-laden structure of modern academe. But what is a conflict of interest? That's open to interpretation, MIT explains in laying out the rules for good behavior by its faculty:

> There are situations sufficiently complex that judgments may differ as to whether there is or may be a conflict of interest, and individuals may inadvertently place themselves in situations where conflicts exist. Accordingly, anyone with a personal interest that may have the potential for conflict with the interests or welfare of the Institute should seek advice and guidance by reviewing the circumstances with the department head, center or laboratory director (who, in the case of sponsored research, should consult with the Director of the Office of Sponsored Programs, or other such person as may be designated by the President). The Faculty Committee on Outside Professional Activities is available for consultation in doubtful situations or those of unusual complexity.[20]

However, strict as they seem, the regulations struggle against academe's deeply embedded laissez-faire traditions. As former Harvard president Derek Bok observed:

> The university strikes many critics as a kind of anarchy, ill-suited for any purpose other than securing the comfort and

convenience of the tenured professors. Officials of the university have very little authority over their senior faculty. The latter have virtually complete license to do as they choose, thanks
to the security of tenure buttressed by the safeguards of academic freedom. Since it is difficult to monitor closely the work
of highly educated professionals, faculty members can travel
more than the university rules allow or remain at home tending their garden or enjoying their hobbies without much fear
of detection. So long as they meet their scheduled classes and
refrain from criminal acts, they can stay happily in their jobs
until they retire.[21]

Even after allowing for a whimsical bent on Bok's part, the reality
is that university faculties are not amenable to regimentation and are
not easily monitored in their comings and goings and outside dealings. At the University of Pennsylvania, where painful memories of the
Gelsinger case persist, I asked Perry Molinoff, the vice provost for research: "Is it your impression that things are more or less under control, or could there be a lot of stuff going on that you might not hear
about until it's too late?"

"I don't think there's an easy answer to that," he replied. "It depends
on what you mean by 'under control.' Do I think that 95 or 99 percent
of the time we're doing it correctly now? I think probably, yeah. But
there are at least four [cases] on my desk right now, which says we're
not 100 percent." The four, he emphasized, were under investigation,
with the outcomes not yet determined.[22]

The rules and regulations for assuring purity necessarily extend
to the finances and commercial ties of family members, leading to
complaints of bureaucratic interference and invasion of privacy from
professors eager to get on with their research. University administrators complain that, in trying to court industry, strict adherence to the
regulations burdens them with a reputation for being difficult to do
business with. Regis Kelly, former executive vice chancellor of the University of California, San Francisco, lamented that the school has such
a reputation. It's derived, he said, from a triad of regulations laid down
by the federal and state governments and the University of California
to protect patients and university property and to assure the university
a proper share if the unlikely occurs and a blockbuster product comes
from commercially sponsored or collaborative research. Kelly cited difficulties arising from an increasingly troublesome legal hurdle for cooperation between institutions: material transfer agreements (MTAs),

which govern rights, obligations, and responsibilities when one research organization provides another with reagents or other substances for research. Commercial organizations can deal as they please, but for academic researchers, the mores of science—and the rules of federal granting agencies—call for all to cooperate in the good cause of advancing scientific understanding and human well-being. However, the possibility of a commercial blockbuster can never be ruled out. With MTAs, the frequent sticking point becomes who's entitled to what if the recipient strikes gold with the transferred material. When he was vice chancellor, Kelly recalled:

> I started trying to do something about this. I tilted at windmills. I would have faculty members coming to me in tears, saying they're all set for an experiment, they have their grants approved, the company was ready to give them the drug, but our tech-transfer office wouldn't let them have the drug, because we couldn't agree on a policy of tech transfer. People were saying the chances, the probability of this having any financial outcome is so minuscule that it's worth taking a risk. So you could do a balanced risk assessment. Unfortunately, it's such a soul-destroying job to deal with material transfer agreements that nobody with a PhD wants to go near them.

To facilitate research, the NIH, with its customary good intentions, directs recipients of its money to share materials. But the NIH has little regulatory authority and even less spirit for policing the practices and policies of the scattered thousands of scientists using its money. "Yes," Kelly pointed out, "they actually came out with a standard form which everybody is supposed to use. Nobody uses it. Not even NIH. The story on the street here," he laughingly explained, "is this is the way the lawyers get business. This is a wonderful thing. A huge amount of legal work involved here. Do you know any lawyers who are going to do anything to reduce the amount of their work?"[23]

By the late 1990s, with academic biomedical research and entrepreneurial deals running strong, the inadequacies of the regulatory system had been extensively documented and were well known within the biomedical research community. But the statistics and success stories of the Association of University Technology Managers told another tale. Issued annually, and uncritically relayed to the public by local and national news media, the numbers and the accompanying narrative were indeed impressive: Universities were spinning out wonderful inventions from their publicly supported laboratories and were reaping income that financed even more scientific advances and rewarded deserving scientists. With so much going so well in the transfer of technology from campus to corporation, there was no consensus about the need or urgency for corrective measures. Besides, as we've seen, ethical safeguards were in place or were being adopted by many universities. Soon, however, several disturbing episodes undermined the long-standing complacency and forced the biomedical research community to face up to ethical failings in the course of human experimentation and other aspects of biomedical research.

7   A New Regime

At the end of the day, the lapses in ethics, the failure to have strong standards, this is about the same thing it's always about. It's about ambition and
greed, not simply in the [pharmaceutical] industry but in my business, too,
in the scientific enterprise, in the universities, without the proper leadership from us. . . . This is about ambitious young scientists who not only
simply want to get tenure but want to win Nobel prizes, as well as it is about
people that saw money to be made in the health-care system that had a huge
amount of money. At the end of the day we have to clean this up. We don't
have any other kind of relationship other than this very fragile relationship
between the industry and basic science in the universities. . . . We just have
to get our ethics straight, and it's not rocket science about transparency and
disclosure and making sure you've got your facts right.

Donna Shalala, president, University of Miami, formerly president
of the University of Wisconsin and secretary of the
U.S. Department of Health and Human Services[1]

: : :

In the late 1990s, the understaffed and listless Office for
Protection from Research Risks, criticized by the General
Accounting Office and patient-advocacy organizations
for poorly performing the role in its title, unexpectedly
rose up and exercised its regulatory power. In March
1999 OPRR ordered a suspension of some four hundred
human trials supported by the NIH at the Veterans Administration health-care complex in Los Angeles. The
VA facilities, affiliated with the University of California,

Los Angeles, included the largest of the VA's 173 hospitals. The VA itself expanded the suspension to animal and other research outside the OPRR order, bringing to a halt a total of about a thousand projects. The drastic action was taken because of "serious deficiencies in human subject protection," OPRR's director of Human Subjects Protection, J. Thomas Puglisi, later told a congressional hearing. Since 1993, he testified, OPRR had repeatedly cited the Los Angeles VA facilities for violations of basic human-protection requirements. The failings included IRB meetings "without a valid quorum" and without the presence of the required nonscientist member and appropriate specialists for reviewing psychiatric studies.[2] Patients were not given sufficient information for providing informed consent to serve as experimental subjects. For six years, repeated but unfulfilled pledges of improvement had warded off a crackdown for serial violations of government-mandated regulations for protecting volunteer patients in medical experiments. A pattern of warning, promise of reform, nonperformance, and nonenforcement had long been typical of the system. But no longer. The shutdown at the VA center was brief but humiliating, and was lifted only when steps toward compliance were clearly evident. Suddenly militant, OPRR issued ominous warnings about further punitive steps. Careful documentation of compliance with IRB regulations was deemed essential. In a talk to the AAMC's executive council, the core leadership of academic biomedical research, OPRR director Gary Ellis wryly stated a metaphysical law of regulatory enforcement: "If it wasn't documented, it didn't happen."[3]

The VA shutdown should have alerted the biomedical chieftains to the arrival of a new era, but it wasn't sufficient to cut into the routinely overcrowded schedules of academic executives and the administrative inertia of their organizations. After all, many of them had taken steps of one sort or another to install the patient safeguards demanded by Washington, while wrestling with the complexities of commercialization on campus and trying to bring in corporate money without trading away too much in academic values. Bruce Alberts, president of the National Academy of Sciences, reflected on a little-recognized impediment to academic reform when we spoke in 2004:

> It's just hard to change a system when everybody in it is sort of running at full speed and has got no time. The faculty I see at universities are busier than I am. Just faculty members. E-mail's made everything worse and faster and faster. Everybody is exhausted and nobody's got time for the long term.[4]

Too busy is not a justification, but it was a factor in the sluggish response to the repeated warnings about IRB performance and patient protection. Unlike military or corporate organizations, universities do not have snap-to-attention chains of command. Power and influence are widely distributed, starting with the chief executive on campus and extending to trustees, faculty, alumni, state legislators and other politicians in public universities, and, to some extent, even the students. The toppling of Lawrence Summers from the powerful presidency of Harvard in 2006 demonstrated that no academic leader is immune to campus opinion. Physically and administratively, medical schools and associated hospitals—the focal point of IRB regulations—are usually separate from the main campus and largely self-governing, with their own dispersion of authority. Given the catalog of goals, problems, and crises that intractably confront higher education, the IRB shortcomings didn't rank high on the academic agenda. Despite the constant sermonizing about the importance of paperwork, especially informed-consent documentation, clinical trials proceeded safely, with very rare exceptions. There was so much else to contend with in a big, modern university. IRBs commanded little attention at the top.

With the VA shutdown, Ellis signaled his determination to enforce the rules. But the message failed to penetrate. The Los Angeles VA center was not a mainstream, big-league biomedical research institution. The premier institutions of biomedical research did not consider it a peer. The political leadership of the biomedical research community was mainly drawn from the big, well-known schools. Lesser institutions were not acutely tuned into Washington's changing politics of patient protection. As is often the case with repeated warnings of dangers that have long failed to materialize, the surprising crackdown on the VA appeared to be a fluke happening and did not register as an omen of broad change. However, the new insistence on adherence to IRB requirements became emphatically clear a few months later with a shutdown order that boomed throughout universities and academic medical centers across the country.

The Crackdown

In May 1999, to the astonishment of the biomedical research community, OPRR ordered a total shutdown of all federally financed research at one of the nation's largest and most prestigious research institutions, the Duke University Medical Center, recipient at the time of about $175 million a year in NIH funds. The drastic step extended to two thousand research projects and lasted four days, thus disrupting or endangering laboratory experiments that required frequent tending as well as clinical trials. OPRR did not claim that research participants had been harmed at Duke, nor was there any indication of imminent danger to them. Rather, it based its action on Duke's unresponsiveness to its repeated expressions of concern about inadequate patient safeguards, poor record keeping, and a scent of conflict of interest. In a visit to Duke several months before the shutdown order, OPRR investigators observed that the IRB membership included Duke's director of grants and contracts. In that job, this campus official was responsible for bringing money into the university and thus, perhaps, was not acutely sensitive to patient-protection violations that might keep it out. Was this a conflict of interest? Nowhere was it written that such an official was ineligible for IRB membership. But the newly militant OPRR was suspicious. Its investigators also raised concerns about record keeping and whether the required quorum was present for IRB deliberations and decisions.[5] These were congenital, enduring deficiencies in the IRB system, as stated in critical reports by the General Accounting Office and others. In the past OPRR either ignored such failings or if it cited them, peacefully departed upon receiving assurances that they would be corrected. But now the agency was on the warpath.

Duke quickly responded to the shutdown with a burst of *mea culpa* correctives, first by establishing a task force to review its procedures for protecting patients in experiments. The single IRB that Duke previously relied on to monitor all such experiments was replaced by four IRBs, each focused on specific areas of research. The supporting staff for the IRB functions was increased from two to eleven, a burdensome expansion in the tightly budgeted finances of academic science. A training program in IRB requirements and the history of medical ethics was established for clinical researchers. A basic tenet of patient protection was stated by Jeremy Sugarman, then director of Duke's Center for the Study of Medical Ethics and Humanities: "Duke has to take the subject of human research on quickly, seriously, and with the necessary resources."[6]

Later in 1999 came the Gelsinger case at Penn, with its repulsive circumstances: an altruistic teenage volunteer dying at a prestigious university in a mismanaged medical experiment steeped in big money and conflicted interests. The youth's death received extensive press coverage that suggested a chilly indifference among scientists to the well-being of the good-hearted volunteers who serve as guinea pigs. The sad episode smacked of failed expertise and innocence betrayed.

Primal fears of rogue science were stoked by experiments gone fatally awry, though the incidence of serious harm inflicted in clinical trials is apparently extremely rare. By one recent account, the estimated 50,000 clinical trials annually conducted in the United States typically result in 17,200 "adverse events," 800 "serious adverse events," and deaths in 1 out of 10,000 patients.[7] Causation is often uncertain because, in addition to healthy volunteers, clinical trials include severely ill people, motivated by altruism or hopes for therapeutic benefit.

However, while the occurrence of deaths in clinical trials appears to be reliably reported, there seems to be limited interest in determining the cause. A study of nearly 900 papers published in leading medical journals in 1994 found deaths reported in 223 of the papers, but autopsies were reported in only 29 papers and "many publications contained no information about the causes of death."[8] Autopsies have generally declined in modern medical practice, largely because of cost restraints. With little incentive to plumb the complexities of causation of death, and limited or nonexistent budgets for that purpose, clinical researchers focused their time and resources on other matters. The frequency of fatalities is in dispute, with some critics contending that the deaths and injuries in experimental settings are grossly underreported—to the extent of "thousands of deaths and tens of thousands of adverse events," according to Adil Shamoo, a bioethicist at the University of Maryland.[9] There's scant support for Shamoo's allegations. But whatever the true numbers for deaths and injuries attributable to clinical trials, in the peculiarities of editors' news judgment, such events, like the occasional beached whale or runaway bride, are newsworthy, because they are infrequent and, in a well-ordered world, are not supposed to happen. In contrast, sparse media attention is given to the horrendous death toll attributed to medical errors in hospitals, estimated as high as 98,000 per year, according to the Institute of Medicine.[10]

The differing degrees of attention merit our notice because they reflect the public's esteem for the sciences, particularly medical science, which is regarded as a beneficent enterprise, despite the undertones of concern about ethical shortcomings. Deaths in hospitals are frequent and familiar, from avoidable as well as unavoidable causes, and are rarely reported outside of the obituary columns. The news media's and the public's concerned attention to deaths in experimental settings reflects acceptance of science's own avowals of high ethical standards, great skill, and beneficent intentions. Opinion surveys steadily confirm that science stands high in the public esteem, with, for example, 86 percent of respondents in one survey typically agreeing that "science and technology are making our lives healthier, easier, and more comfortable." [11]

Fearing Loss of Federal Support

A fall from grace by scientists evokes gasps and opprobrium in the popular press and among the public and, in turn, raises a baseless though plausible fear in the halls of science—that a loss of public trust will lead to a loss of public money for research. So it might seem, but it isn't so. Nonetheless, the leaders of science believe it is so and react to that belief—to the benefit of upgrading ethical performance in their profession, or at least earnestly talking about it. Through the 1980s and 1990s, though many cases of scientific fraud and other wrong-doings in research came to public attention, government support for science across virtually all disciplines nonetheless rose briskly. The medical sciences did especially well, rising from $3.7 billion in 1980 to nearly $18 billion in 2000.[12] At work during those decades were the stable motivations for government spending on research: initially, Cold War anxieties, which were then succeeded by the ever-present hopes for medical cures and economic growth.

We must conclude that no tight correlation exists between misdeeds in science and congressional appropriations for science. The scientists' fear of a backlash from wrongdoing in research is attributable to anxiety rather than the workings of the political system. Politicians have never yet cut budgets for research in retaliation for scientific delinquency. In fact, the short stretch of years in which Congress doubled the NIH budget, 1998–2003, included some of the most disturbing failures in protection of volunteers in medical experiments. Still, the budget juggernaut rolled on. But dependent as they are on the NIH's money, the managers of academic medical research easily come down with fiscal jitters. Their calls for reform may have been humanitarian in origin but also were explicitly tied to material practicality: If we don't shape up, we'll lose money. The enforcers of patient-protection regulations were at last getting the attention of biomedical researchers and their administrators. Propelled by a fear-ridden misunderstanding of the workings of the political system, the biomedical-research enterprise awakened to what it perceived as danger in neglecting the IRB rules.

Several more shutdowns were ordered by the invigorated OPRR: the University of Illinois at Chicago in August 1999 and Virginia Commonwealth University in January 2000. Their shaming, along with the public humiliations of Penn and Duke, seemed at last to produce salutary effects. Among researchers, the targeted universities became worrisome symbols of institutional reputations damaged by inattention to the basic medical and bureaucratic requirements of ethically conducted research on humans. "It put us on guard to say, were we looking at ourselves well enough, do we have our systems in good enough shape?" Steven A. Goldstein, associate dean for research and graduate studies, at the University of Michigan Medical School, told me. "So we looked hard and we said we can improve a lot of our systems, and we did that. As a result of that, we've done very well. We've not had any huge problems, and I think that's healthy." [13]

Meanwhile, in the wake of the Penn and Duke disasters, the two leading Washington-based organizations representing academic science, the Association of American Medical Colleges and the Association of American Universities, were aroused from their chronic languor by the wave of federal crackdowns on faulty ethical performance. The direct stimulant for their awakening was fear that biomedical misdeeds might derail Congress's on-time progress toward fulfilling its commitment to double the NIH budget between 1998 and 2003. The NIH, heavily mortgaged with prior multiyear grants, had little to spare for new obligations. The budget-doubling plan provided the best kind of money for research, "new money," as it's known in academic circles, which allows new programs to be launched in the universities. In customary fashion, the concerns within the two university associations led to the creation of committees to conduct still further inquiries into the already well-masticated ethical issues arising from research and to recommend improvements. The AAMC, with its 125 medical-school members, and the AAU, comprising sixty American research universities and two in Canada, undertook separate studies of conflicts of interest and protection of human subjects in academic research. The announcement of the studies, which would be long in progress, provided a rationale for opposing immediate tightening of federal regulations.

A New Cop on the Beat

OPRR director Gary Ellis's exercise of regulatory muscle ignited a strong reaction in politically well-connected biomedical circles. If only to shake off damaging publicity and budgetary retribution, the need for stricter ethical safeguards was now accepted by academic leaders, and they were moving toward that goal, though slowly. But Ellis, tolerable as a cautious bureaucrat, had now become an intolerable rampaging regulator, with many vulnerable targets still untouched by his investigators. The universities in his sights were not politically help-less or naive. The big schools maintain offices in Washington, while collectively various sectors of academe are represented by such major organizations as the AAMC and the AAU. Dependent on government money and sensitive to federal regulation, science and higher education know their way around Capitol Hill and regularly make their wishes and fears known to hometown congressmen and senators, who are ever responsive to opportunities to oblige their constituents. In the ex-ecutive branch of government, department heads seek cordial relations with the legislators who control their budgets and legislative author-ity. Capitol Hill and the Department of Health and Human Services shared discomfort about the Ellis problem.

Moreover, as home base for OPRR, the NIH was uneasy with the dual tasks of financing research and monitoring its ethical status in universities and in its own facilities. The combined roles of patron and cop looked like a conflict of interest. Several years earlier, concern about conflicting roles had led to the relocation of another watchdog agency, the Office of Scientific Integrity, which was responsible for po-licing fraud in research. Originally based in the NIH, the office was moved to its parent department, Health and Human Services, to put some bureaucratic distance between it and the NIH. Later an advisory panel recommended a similar shift for OPRR, on the same grounds of conflicted roles.[14] In June 2000, with the Gelsinger and Duke episodes freshly reverberating in academic and government circles, Ellis's boss, HHS secretary Donna Shalala—a former chancellor of one of the larg-est research universities, the University of Wisconsin–Madison—or-dered a bureaucratic reshuffling of responsibility for IRBs and human experimentation. Shalala abruptly abolished the NIH-based Office for Protection from Research Risks. In its place, she established a similar entity with a slightly different name, the Office for Human Research Protections (OHRP). Ellis was out.

With the name change and removal of Ellis came a new location, a new chief, and department-wide scope for overseeing patient pro-tection in experiments financed by all HHS agencies, not just by the NIH. The newly created OHRP was attached to Secretary Shalala's departmental office, where it came under the federal government's top health official, the assistant secretary for health—a significant upgrade in the federal hierarchy, denoting higher political interest. Named as the first director of OHRP was Dr. Greg Koski, a Harvard Medical School professor. As director of human research affairs for Partners HealthCare system, Koski oversaw patient protection for research conducted at the Harvard-affiliated Massachusetts General Hospital, Brigham and Women's Hospital, and the Dana-Farber Cancer Institute. In expressing her expectations for higher ethical practices in medical experimentation, Shalala invoked the instantly recognized symbol of the patient-protection system gone wrong: "The tragic death of Jesse Gelsinger," she stated, "focused national attention on the inadequacies in the current system of protection for human subjects." To accelerate progress toward reform, Shalala announced, her department would seek legislation authorizing fines of $250,000 per investigator and $1 million per research institution for "violations of informed consent and other important research practices." [15]

The specter of big fines was disturbing for academic administrators and researchers, who regarded themselves as honorable professionals engaged in difficult work that was essential for the well-being of the American people. On the other hand, the removal of Ellis, his replacement with an academic physician, and the elevation to departmental headquarters raised expectations of more sensitive treatment of universities struggling with the intertwined complexities of modern academic research: hauling in sorely needed research money from government and industry, avoiding or neutralizing conflicts of interest, and recruiting the requisite human volunteers for experimentation, while upholding high ethical standards, and filling in the paperwork demanded by government overseers. The Koski regime was both surprising and brief.

A System "Out of Control"?

Koski arrived in Washington hopeful about raising the ethical performance of academic science, but also acknowledging that research and commercialization had become so extensively intertwined that "management" of troublesome relationships, rather than their elimination, might be the only realistic goal. On August 15–16, 2000, following announcement of his appointment but two weeks before he was to take office, he was in the audience at a conference at the NIH on "Human Subject Protection and Financial Conflicts of Interest." IRBs were not required to monitor financial conflicts, though some universities assigned them that task. But the pairing of financial interests and protection of human subjects was rising on the agenda of biomedical policy concerns as researchers and universities became entangled financially with the biotechnology and pharmaceutical industries and the start-up companies of their own professors. The urgency and amorphous boundaries of the topic were evident from the disparate cast now heavily involved in the conduct, management, and profit-seeking of twenty-first-century biomedical research. In attendance were some seven hundred biomedical researchers, administrators, bioethicists, bio-business executives, lawyers, and others from throughout the country, meeting in Washington in prime summer vacation season for two long days of deliberations. Invited to address the audience, OHRP director-designate Koski spoke in an extemporaneous, informal fashion, expressing ethical concerns and ambitious plans, as well as a sense of real politik concerning the presence of commercial money in academic biomedical research:

> We know that money talks. The drug companies will tell us that money talks because it is the single most effective way to encourage human research subjects to participate in research. It is the single most effective way to encourage individual investigators to complete enrollment goals for studies. So, money clearly talks. . . . I would call the conflicts of interest [in clinical research] pervasive. Indeed, in an enterprise where we have embodied in single individuals dual conflicting roles, the physician-investigator, the patient-subject, there are going to exist conflicts of interest that simply are not something that we can eliminate. They are inherent. They are intrinsic and unavoidable to the research process. . . . But one of our challenges continues to be how to manage those conflicts that we cannot eliminate. So that as in so many other complex environments, where it would be nice to have an ideal, clean situation, I don't believe we are truly going to see that in this domain. So that the continued emphasis on management of conflicts of interest in an effective manner, when elimination of the conflicts, which would be our first goal, is simply not possible, is going to be one of our big challenges. The truth of the matter is . . . some of the research that in my mind poses some of our greatest challenges and risks is being done outside the academic setting right now. It is being done in private physicians' practices. It is being done in the private research centers . . . that I think admittedly may not fall under the same kind of administrative oversight and public scrutiny that is essential. . . . Many have pointed out that we currently have multiple sets of regulations regarding conflicts of interest that have been promulgated by different agencies under different regulatory codes. I don't believe this is a situation that we can continue to allow to prevent us from achieving the goal of having consistency across

the board. First of all, it is quite clear [from discussions at the conference] that conflicts of interest are very real. They are very serious and they are a threat to our entire endeavor. These conflicts have certainly intensified over the last two decades and certainly during the last five years, the system may have gotten entirely out of control. There is a need to very immediately, at least to begin, to get the system back into some kind of control.

Koski also interjected caustic remarks about the recent conflict-of-interest debates at his own institution, Harvard:

If I could digress for one moment, I simply have to say that as the faculty and administrators at Harvard Medical School considered the possibility of revising their conflict-of-interest policies . . . I read accounts that indicated that one of the concerns was that stringent policies on conflict of interest would make it difficult to retain and recruit faculty. It is almost hard for me to say that, but I think that it is a sad commentary on the status of science and academics to say that stringent policies on conflicts of interest to protect the integrity of science and the well-being of research subjects would be an impediment to recruiting and retaining faculty.[16]

The views expressed by Koski would once have offended the biomedical establishment and provoked it into self-righteous, indignant rejoinder. But within the leadership, fears were growing of a political and financial backlash from reports of misdeeds in medical research. The peril was soon candidly addressed by AAMC president Jordan Cohen, head of the medical school lobby, who warned his biomedical-research constituents that the appearance of greed and indifference to patient safety could have costly consequences. In October 2000, shortly after Koski took office, Cohen gave an address, "Trust Us to Make a Difference," in which he departed from soporific association talk and bluntly warned that failures of patient protection, real or merely perceived, might undermine the doubling plan for the NIH budget:

Are we supposed to wait for irrefutable evidence that some patient died because an investigator with a financial conflict of interest was so blinded by greed that he or she failed to do all that could be done to prevent the death? Or does the under-

standable public *perception* that such a nightmare is possible
necessitate public action now? . . . I needn't belabor what's at
stake here. Public trust is what fuels public support for medical
research. Imagine how willing the public and their representatives would be . . . to double the NIH budget if serious concern
were widespread that financial interests on the part of university investigators biased their research, warping the results in
their favor at the expense of objectivity. . . . And then imagine
how proud you would be to be part of an enterprise viewed
by the public with skepticism rather than with the admiration
and esteem you now enjoy. So whether or not external financial interests have resulted in an *actual* degrading of sound
scientific practice . . . we risk great peril if we fail to respond to
the growing perception that financial conflicts of interest have
gotten out of control. A perception, I would remind you, that
is shared by Congress and by the Department of Health and
Human Services.[17]

Politics Intrudes

Though the necessity for reform had at last penetrated the conscious-
ness of the biomedical leadership, the old problems of patient protection and conflicts of interest still persisted. Koski, coming from an
academic background, looked like a good prospect for dealing with
them sensitively, in a cooperative, nonabrasive manner. But the new
man at the helm of the newly established Office for Human Research
Protections was soon preoccupied with his own shaky political plight.
Koski arrived in his new job in September 2000 as an appointee of
a lame-duck Democratic administration. It's likely that he would
have fared better if Al Gore had won the presidency in the November
2000 election. As a young congressman, Gore had presided at House
committee inquiries into medical ethics and human experimentation.
He was personally acquainted with many scientists, and, rare among
politicians, appeared genuinely interested in science and technology.
Koski's office was too far down in the federal hierarchy to engage close
presidential attention, but a Gore administration would have provided
a friendly, supportive milieu for improved policing of human experimentation. Democratic administrations were comfortable with wield-
ing regulatory power, but were also in harmony with universities,
politically friendly places from which many Democratic appointees for
government posts were drawn.

However, in January 2001, four months after Koski took office, George W. Bush was sworn in as president, bringing in an administration proudly and aggressively hostile to government regulation. The new administration, disinclined to butt heads with academe on the obscure subject of protection of volunteers in medical experiments, was wary of adopting stiffer regulations or snooping on academic practices. The Republican-controlled Congress ignored Shalala's proposal for fines, as did her Republican successor as HHS secretary, former governor Tommy Thompson of Wisconsin.

Meanwhile, as Koski settled into his new job in the unfamiliar territory of Washington, and as a holdover from the Clinton administration, the General Accounting Office produced another report about IRB failings, further detailing their flimsy grip on the management of conflicts of interest and protection of experimental subjects. Confusion about the rules, or indifference to them, was evident from the GAO's sampling of five major universities that received large amounts of NIH money: UCLA; University of North Carolina at Chapel Hill; University of Washington, Seattle; Washington University, in St. Louis; and Yale University. No wrongdoing or endangerment of patients was reported by the GAO, which did not disclose which of the universities fell short in any respect. But it was clear from the report that the five universities were not acutely concerned about conflicts of interest and patient protection, and it could be assumed that they were representative of many other schools. The GAO stated:

> All five universities had difficulty providing basic data on investigators' financial conflicts of interest in clinical research involving human subjects. The universities generally acknowledged a need for better coordination of information about investigators' financial relationships, and several of the universities told us they were developing mechanisms to do so. . . . One university . . . mistakenly assumed it needed to report only the financial conflicts of interest that could not be managed; therefore, if it had eliminated a conflict of interest, it did not report it. . . . At two universities . . . investigators had to disclose [their financial] interests to their study subjects. . . . The other three universities decided on a case-by-case basis whether investigators would be required to disclose financial interests on the consent form. . . . Of 111 investigators at four of the universities we visited who had significant financial relationships with industry in 2000, only 3 voluntarily divested their interests;

none were told to divest by their universities. . . . None of the five universities had formal processes for verifying that individuals fully disclosed their financial interests.[18]

The Other Conflict of Interest

In January 2001, in his first major step as head of OHRP, Koski published a "draft interim guidance" concerning the old nettlesome issue of conflicts of interest in government-financed research. The draft, posted on the Internet, was intended to update well-intentioned but soft regulations issued in 1995. The few intervening years had brought many exposures and criticisms of unwholesome commercialism in academe. In accordance with standard federal procedures, Koski's draft was published as a discussion piece for comments by interested parties. It could have no regulatory bite until comments had been received and, presumably, given fair consideration, prior to a final version taking effect. Included in the document was a segment of the conflict issue that had long been untouched, left on the sidelines as simply too difficult to address, let alone fix—*institutional* conflicts of interest. These were different from *individual* conflicts of interest. Troublesome as they were, individuals with conflicted situations were familiar subjects in ethical deliberations and policymaking, and most universities knew of various ways to deal with their conflicts, even if they chose not to, which was often the outcome.

But institutional conflicts of interest presented a tangle of problems that had heretofore been ignored. Many of the big research universities were financially permeated with a variety of entrepreneurial activities linked to their laboratories, such as investments in their professors' start-up companies, clinical trials of drugs developed on campus and licensed to outside firms in which the university held stock, and endowment holdings in pharmaceutical companies that provided unrestricted philanthropic funds or financed specific research projects on campus. Not uncommonly, medical specialists at prestigious research centers endorsed products manufactured by firms in which they and their institutions held financial stakes. For example, in 2005 the *Wall Street Journal* revealed that the CEO of the Cleveland Clinic, Delos "Toby" Cosgrove, was a leading cheerleader for a heart-lung machine manufactured by a firm in which he and the clinic held financial interests. Following tests at the clinic, Cosgrove reported that "the results look very encouraging and exceeded our expectations," according to the *Journal* report.[19] In several publications and professional talks, the financial connections were not disclosed. The Bayh-Dole Act not only permitted but required universities to seek the commercialization of their scientific output. But beyond the legislation's singular focus on patent rights, many incestuous transactions and relationships flourished. Uncertainty hung over the issues of who was responsible for looking out for trouble in these deals and who held authority to block tawdry situations. Koski's document openly addressed the growth and potential perils of commercial enterprise at the institutional level, thus taking the discussion of conflict of interest into sensitive territory. Like a closeted scandal in the family, these were matters best kept from outsiders. But Koski, the Harvard professor on leave in Washington, nonetheless waded in, asserting in the proposed guidance document that safeguards were needed to prevent universities from chasing profits at the expense of scientific integrity and harm to volunteers in medical experiments. That his warning foretold the circumstances of the Cleveland Clinic case five years later should not be attributed to prescience. Many such deals both preceded and followed Koski's service in Washington, and their existence was well known among researchers and government regulators. Without genuflections to academe's virtue, the guideline stated:

> Increasingly, academic institutions and corporate entities are entering into agreements that are mutually beneficial, and which may also bring the institution's interests into direct conflict with those of research participants. For example, an institution may accept a principal equity interest in a biotechnology company as part of a cooperative endeavor to develop a new medical device. Clearly, in such a situation, both the institution and the corporate partner would stand to gain financially if the device proves to be safe and effective. Accordingly, the institution should carefully consider whether a clinical trial to evaluate safety and efficacy should be performed at that site, and if it should, what special protections would be needed. The financial interest of the institution in the successful outcome of the trial could directly influence the conduct of the trial, including enrollment of subjects, adverse event reporting or evaluation of efficacy data. In such cases, the integrity of the research, as well as the integrity of the institution and its corporate partner, and the well-being of the research participants, may best be protected by having the clinical trial performed and evaluated by independent investigators at sites that do not have a financial stake in the outcome of the trial, or carried

out at the institution but with special safeguards to maximally protect the scientific integrity of the research as well as the integrity of the study and the research participants.[20]

Koski's document mainly called for closer attention to individual and institutional financial conflicts, and their disclosure and management—hardly a radical or onerous proposition, given how far government and academe had come in recognizing the failings of the ongoing system. The term "draft interim guidance" suggested a cautious, tentative, tiptoe approach. But like a virulent microbe setting off an immune reaction, the document aroused opposition in important places. An official of the National Science Foundation cautioned that "guidelines" were often mistaken for regulations, thus suggesting risk in hasty action. The NSF was widely regarded as a well-run agency that managed to be sensitive to the needs of universities while strictly observant of federal regulations; its opinion counted. Objections were raised by the Federation of American Societies for Experimental Biology, which at the time represented sixty thousand scientists in twenty-one scientific societies. FASEB, known as the Washington voice of bench scientists— the frontline workers of science—declared its support for "federal guidelines, not regulations, to assist universities and other organizations in designing rigorous and locally appropriate policies." FASEB's position, stated in a letter from its top two officers, frankly argued that "in order to encourage the translation of fundamental discoveries into novel modalities of patient care, some degree of financial conflict of interest is to be expected."[21] FASEB and other organizations pointed out that the AAMC committee was still at work on its study of financial conflict of interest, though the AAU committee had already delivered a report on protecting volunteers in experimental research. These organizations were usually concerned with fending off federal regulation, invoking the customary assurances of academe's integrity and dedication to the public interest as safeguards against misbehavior. But they also feared that science in universities was acquiring a bad name with the public, and that the appearance of complacency on their part could be detrimental. The party line was under pressure.

Delivered in June 2000, the AAU's *Report on University Protections of Human Beings Who Are the Subjects of Research* was a vapid production, steeped in common-denominator platitudes. "The Task Force," it declared, "urges prompt attention to strengthen human subjects protection to . . . ensure that the highest standards are being followed in protecting the rights and welfare of human beings." Better protection of human subjects, the AAU explained, would "reduce the likelihood of inducing changes to laws and regulations that might bring other unforeseen consequences." [22] Scant attention was paid to these pronouncements, deservedly so. But the AAU had another report in progress, dealing, in part, with institutional conflicts of interest. In the division of labor between the AAU and the medical-school association, the AAU focused mainly on the National Science Foundation, which financed academic research outside the medical sciences, while the AAMC concentrated on the NIH, the main bankroll for research in its member institutions. Both academic organizations, joined by others, called for withdrawal of Koski's draft guidelines. AAMC vice president David Korn said the guidelines were "quite premature," adding: "I think it is necessary to address these issues, but I don't think the government has any great wisdom to [offer]. We don't even know how to define an institutional conflict of interest." Koski replied: "We haven't issued any guidance yet and you can't withdraw something that hasn't been issued." [23] Still to come was the AAMC report, which was moving along very slowly. While it was gestating, the biomedical community again assured the public of its high integrity.

In June 2000 three hundred universities, research centers, professional associations, and hospitals issued a statement titled "Clinical Research: A Reaffirmation of Trust between Medical Science and the Public." In an accompanying press release, AAMC president Jordan Cohen was quoted as saying, "The academic medical community is committed to the health and welfare of all individuals who participate in clinical research." Without explicitly referring to the shutdown orders at universities, Cohen acknowledged that "recent well-publicized events have shaken the all-important foundation of trust between researcher and patient. The AAMC and its members will take the necessary actions to rebuild this foundation that is so crucial to the advancement of science and the delivery of quality health care." Similar sentiments were expressed in the statement by the head of the National Health Council, which improbably claimed to represent "more than 100 million people with chronic diseases and/or disabilities."

The statement of the three hundred institutions pledged strict adherence to the highest ethical principles, including "that patients are informed of any reasonably foreseeable risks and benefits of participating in the research activity." And it reaffirmed "the central importance of, and adherence to, the procedures mandated by federal human subjects regulations, which prescribe a process by which research protocols are reviewed with attention to safety, ethics, and the protection of human participants." The sincerity behind these declarations cannot be doubted. But neither can the denial and self-delusion among institutional leaders who were so convinced of the goodness of their organizations, colleagues, and themselves that reality eluded them. Reform was difficult to achieve, but public-relations balm was easy to deliver. Among the signatories was one of the most prestigious biomedical-research and treatment complexes in the world, the Johns Hopkins School of Medicine and the Johns Hopkins Bayview Medical Center, in Baltimore, Maryland.[24]

A Death at Hopkins

One year after this reaffirmation of trust, a twenty-four-year-old healthy volunteer, Ellen Roche, a laboratory technician in the Johns Hopkins Asthma and Allergy Center, died after being administered two experimental doses of a chemical inhalant, hexamethonium. The fatal experiment, financed by the NIH and conducted at Hopkins, was intended to study how healthy lungs respond to asthma attacks. As stated in a report of Hopkins' internal investigation of the fatal experiment, Roche apparently was motivated to participate by "(1) an altruistic desire to help people with asthma and (2) monetary compensation ($25 for each of the first phase visits and $60 for each of the second phase visits, totaling $365)."[25] The internal investigation revealed a failure of the researchers to review scientific literature indicating risks of hexamethonium. In 2001, at the time of the young woman's death, Hopkins was—and remains—the largest recipient of NIH research money, about $300 million for 2,400 research projects involving some 15,000 human subjects. Koski didn't come to Washington seeking a showdown with one of the world's most renowned biomedical research institutions. But the fatal experiment brought to a head long-lingering issues of Hopkins' noncompliance with IRB regulations. On July 19, 2001, OHRP ordered a suspension of all federally funded research studies approved by review bodies at the Johns Hopkins School of Medicine and its Bayview Research Campus. The Gelsinger case at Penn and the Roche case at Hopkins both involved deaths of volunteers in experiments. But Koski regarded the underlying circumstances at the two institutions as very different from each other, he later told me.

> Penn had a very scandalous case that involved an investigator in particular, but it wasn't so much the institutional review board that was involved in that problem. It was really more of

an investigator and institutional issue. Whereas in the Hopkins case, it was clearly a systemic breakdown, a deficiency of the entire process for protection of human subjects. There were very dramatic differences between the two.[26]

Proud Hopkins, some would say arrogant Hopkins, did not meekly accept the humiliating shutdown decree, which drew widespread media attention. Interviewed on the public television program *The NewsHour* the day after Koski ordered the shutdown, Dr. Edward D. Miller, dean of the Johns Hopkins School of Medicine, loosed an indignant salvo at Koski's Office for Health Research Protections, declaring:

> We find it difficult to understand why a relatively new agency would take these draconian measures in an institution that has cared for thousands of people in clinical trials. We have done clinical trials for over a hundred years here at Hopkins. We have had one death in all of these years in a human, healthy volunteer. For OHRP to take this measure and not understand the consequences on patients that are treated here cannot be understood by me at all.

Miller then flaunted his skill at political navigation, explaining that upon receiving the shutdown order, he appealed directly to Koski's ultimate boss: "I contacted the secretary [of Health and Human Services], Tommy Thompson. He responded to me this morning. We now have a process in place where OHRP and Hopkins are working diligently tonight . . . to have a corrective action plan that will allow full accreditation of Hopkins's research very shortly."[27]

Appearing on the same *NewsHour* program was Dr. Ernest Prentice, associate vice chancellor for academic affairs and regulatory compliance at the University of Nebraska Medical Center. Prentice, who served as an OHRP consultant on compliance issues, explained that

> when OHRP investigates an allegation of noncompliance, they also evaluate the entire program for protection of human subjects. More or less the allegation of noncompliance kind of opens the door for a much wider, systemic evaluation of an institution's program, and that is what OHRP did at Hopkins. . . . So Johns Hopkins' program was shut down not necessarily because of the unfortunate death, but because of the deficiencies identified in that program.

Asked to what extent the problem extended beyond Hopkins, Dr. Prentice expressed a familiar theme: "No, it's not a Hopkins problem; it's a universal problem. We know from a number of different studies that IRBs have been chronically overloaded and under-resourced. So there is a problem out there. I think it's gradually being corrected, but it's going to take some time."[28]

Despite OHRP's assessment of a broadly deficient state of ethical compliance at Hopkins, Koski quickly relented under protests against the wholesale shutdown from Hopkins and its political friends. One day after ordering the suspension of research, OHRP advised Hopkins officials that clinical trials could resume if they decided it was in the best interests of the patients. A few days later, the suspension was entirely lifted upon OHRP's acceptance of a "corrective action plan" submitted by Hopkins. Officials at the university said several months would be required to restore all the suspended research projects.[29]

The swift removal of the shutdown order generated the impression that Koski was reined in by politics. As the Bush administration filled the upper federal ranks with appointees of its choosing, he remained one of the few holdovers from the Clinton administration. A further sign of Koski as a politically isolated outsider came when his departmental superiors disbanded the advisory committee for his office and appointed new members, without consulting him. For aficionados of bureaucratic craft in Washington, this was a milestone event. When shrewdly selected and orchestrated, members of a government advisory committee can amplify an official's influence and power by connecting to existing and potential constituencies and mobilizing their support. Koski later contended that "the advisory process was being manipulated to promote specific ideological viewpoints"—meaning the anti-abortionist strategy to extend patient-protection regulations to embryos and fetuses.[30]

At the start of the new century, the ethical condition of biomedical research in the United States remained, as usual, difficult to read, given the dispersal of health-research activities among several thousand self-contained institutions, ranging from huge to small, each with its own culture, ethical sensitivities, financial pressures, rules, and leadership. Another factor was the continuing dispersal of clinical research to nonacademic sites, including for-profit contract research organizations and third-world locations. Though the NIH emphasized clinical trials as a critical link in the provision of better health care, government support was declining as a share of the total while the pharmaceutical industry increased its support and exercised its influence over the trials.

The NIH and other federal agencies that financed clinical research laid down various guidelines, but did not undertake rigorous surveillance and took drastic punitive steps only in instances of gross violations that could not be ignored. Nonetheless, even as Koski was completing his tour of service in Washington and preparing to return to Harvard, the cause of righteousness in research was nudged forward—not dramatically forward, but significantly so, in comparison to the long run of indifference and dodging that usually prevailed among researchers, administrators, and the organizations that represented academic science in Washington.

Raising the Bar for Rectitude

In December 2001 the snail-paced Association of American Medical Colleges issued the first, and more important, of two reports that were conceived in the aftermath of the Penn and Duke episodes. The death at Hopkins occurred while the report was in preparation, raising the pressure to produce something useful. Titled *Protecting Subjects, Preserving Trust, Promoting Progress,* this was a blue-ribbon document, produced by a carefully selected twenty-eight-member Task Force on Financial Conflicts of Interest in Clinical Research. Mandarins of medical research and education were plentiful in the membership, including, as chairman, William Danforth, chancellor emeritus of Washington University, Saint Louis; and Joseph P. Martin, dean of the medical faculty at Harvard. Reflecting the AAMC's recognition of the importance of public relations, the group included several figures connected to the news industry: Susan Dentzer, of the PBS *NewsHour;* Hedrick Smith, a TV producer formerly with the *New York Times;* and Marvin Kalb, a former TV correspondent and head of the Washington office of Harvard's Joan Shorenstein Center on the Press, Politics and Public Policy. Also on the task force were three senior officials of major biotechnology firms, several attorneys, an ethicist, and the head of the National Breast Cancer Coalition, one of the most effective of the many patient organizations working Capitol Hill for research money. Washington abounds with policy committees handpicked for rank, public prominence, influence, and experience. The task force ranked high in all those respects, as well as in the crucial matter of staff support for its labors, which was headed by AAMC senior vice president David Korn. Committee members, primarily employed elsewhere in demanding jobs, rush to and from meetings. Staff members, employed at headquarters, are there full-time to service the meetings and ghostwrite the report. A former dean of Stanford's medical school, Korn had long sounded alarms about the erosion of ethical standards in academic research. But wary of reform coming from government, Korn advocated reform from within. After the disasters he prophetically feared had actually occurred, he successfully politicked within the AAMC for a strong response. The task force and its report were the result.

The AAMC report was a moral declaration of the need for curtailing the ethical failings of science for sale and an endorsement of safeguards to attain that goal. Compliance would have to be voluntary, as was the case with other policy prescriptions of the medical-school association and its academic counterparts. These organizations have no disciplinary power. But the good name of medical research was now stained by deaths and numerous reports of mercenary dealings by professors with one foot in academe, the other in commerce, to the detriment of patients, the public interest, and the reputation of medical science. The press, the public, and federal enforcers, stirred by concerned scientists and journal editors within the biomedical-research enterprise, now accepted that some things were seriously amiss in the conduct of medical research. The go-go spirit on the commercial frontiers of biotechnology had produced enough misdeeds to sway perceptions. With the shame and embarrassment factors lurking as always, the money-hungry institutions of biomedical research, large and small, were now far more attentive to admonitions about their ethical standards and warnings of the political-financial damage that might ensue from further fatal mishaps in experimental settings and revelations of conflicted dealings.

But we must be careful to distinguish between the willingness to pay attention—which was on the upswing—and a readiness to adopt strong corrective measures that would necessarily entangle academic researchers in even more bureaucratic rules, limits, and paperwork. Of these, as we have seen, many believed they had more than enough. Furthermore, the AAMC report and recommendations for protecting humans in research studies and scientific integrity were confined to individual financial conflicts of interest. The thornier issue of institutional conflicts was set aside for a separate study by the same task force, which took another twenty-two months to produce a report on that topic.

A Higher Standard

For assessing the propriety of individual academic researchers experimenting on human subjects, the task force report advocated the

adoption of a "rebuttable presumption" that the dealings are impermissible if the researcher stands to gain financially. Thus was introduced a challenging legal concept that translates into assumed guilty
until proven innocent, with the burden on the suspect—that is, the researcher—to prove that professional judgment and performance would
not be swayed by the profit motive. In science's slow progress toward
implementing and policing ethical behavior in research, the rebuttable
presumption ranks as a landmark. The AAMC report stated:

> With the welfare of research subjects always of foremost
> concern, an institution should regard all significant financial
> interests in human subjects research as potentially problem-
> atic and, therefore, as requiring close scrutiny. Institutional
> policies should establish the rebuttable presumption that an
> individual who holds a significant financial interest in research
> involving human subjects may not conduct such research. The
> intent is not to suggest that every financial interest jeopardizes
> the welfare of human subjects or the integrity of research, but
> rather to ensure that institutions systematically review any fi-
> nancial interest that might give rise to the perception of conflict
> of interest, and further, that they limit the conduct of human
> subjects research by financially interested individuals to those
> situations in which the circumstances are compelling.[31]

Researchers with special, rare skills might be allowed to participate in human subjects studies despite having a financial interest in
the proceedings, but only after a searching review and under close supervision. Disclosure of financial interests and transparency to render
them visible were deemed indispensable for protection of patients in
experiments and, in general, for the wholesome practice of science.
Targeting academic consultants who shill for the pharmaceutical industry in the guise of engaging in disinterested scientific discourse, the
strictures extended to publications and oral presentations that involved
remuneration from any commercial entity. These mercenaries regularly perform at scientific conferences and at the continuing medical
education courses that practicing physicians are required to take to
retain their medical licenses. Evident in the AAMC document was the
long-lingering fear that ethical misdeeds would lead to loss of public
trust, which would lead to loss of public money. Academic-industrial
collaboration has produced many medically beneficial results, the report emphasized, but it also cautioned that "the public's extraordinary

support of academic biomedical research will remain critically dependent upon public confidence and trust that are especially vulnerable in research involving human subjects. This is the reality, and it must be appreciated by industry as much as by academe if their interactions are to thrive." [32]

Only one member of the task force declined to endorse the report, reflecting the tight linkage of academic science and the biotech industry. A footnote on the page listing the twenty-eight task force members states that Susan Hellman, the chief medical officer of Genentech, "declines to endorse the report, primarily due to her concern that its recommendations present an impediment to research innovation."

The Other Conflict of Interest

The AAMC and AAU studies of institutional conflicts of interest were reluctantly undertaken and resulted in a mélange of trite observations and inconsequential recommendations. The efforts at reform were stymied because these conflicts arose from commercial relationships and deals that academic administrators found either profitable or promising for their institutions, regardless of the appearance or reality of ethical corner cutting. The AAU report on institutional conflicts of interest, issued in October 2001, explained that universities were experienced in dealing with individual conflicts, but with the institutional type "the focus is on developing policies and principals, since no regulations guide this area. . . . Given the dearth of previous policy making in institutional conflict of interest, the Task Force is cognizant that its efforts are but a first step in developing and institutionalizing processes in this field." [33] Six years later, no further steps were evident.

The AAMC's report on institutional conflicts was released in October 2002. For seasoned readers of the Washington report genre, an early telltale paragraph signaled that the authors took a pass on the issue:

> As an initial response to a problem of remarkable complexity, this report does not provide an exhaustive list of potentially troubling financial interests; nor does it prescribe a comprehensive scheme for the oversight of all institutional relationships with commercial research sponsors. Instead, the report offers a conceptual framework for assessing institutional conflicts of interest and a set of specific recommendations for the oversight of certain financial interests in human subjects research that,

in the view of the AAMC's Task Force, are especially problematic and must therefore receive close scrutiny.[34]

The report prescribed good judgment and transparency in coping
with institutional conflicts of interest and advised its member universities to appoint institutional conflict-of-interest committees to review
dicey situations involving human volunteers in experiments. The
AAMC even gingerly suggested application of the rebuttable presumption concept in conflicted institutional situations. But it backed away
from a direct confrontation with the institutional issue, stating that
"ultimately . . . each institution must determine how best to segregate
human subjects research and investment management functions fully
and reliably within the context of its own organization and governance
structure."

"Hysterical and Overstated?"

Several years after the production of these expressions of piety, an un-
guarded insight into their origin was provided by the key figure in the
process, David Korn. The setting was a public debate on the regulation
of biomedical research held at the conservative, anti-regulatory American Enterprise Institute (AEI), in Washington, D.C.[35] Opposite Korn
was a leading opponent of the thickening web of rules for research, Professor Thomas P. Stossel, of the Harvard Medical School, author of an
anti-regulatory critique published a week earlier in the *New England
Journal of Medicine,* under the title "Regulating Academic-Industrial
Research Relationships—Solving Problems or Stifling Progress?"[36] In
the article Stossel argued that the latter effect was triumphant, stating that "university and governmental rules that prevent wide-ranging
interactions between academic researchers and industry limit creative
and economic opportunities and are a far greater violation of academic
freedom than any documented interference by industry." In his talk
at AEI, Stossel said, "Let's get the bad guys when they do bad things.
Punish them severely. But don't paralyze innovation. . . . All this energy
that is going into sanitizing research and obsessing about conflict of
interest, if we could do a better job . . . of getting academic technology matched appropriately with industry, what a better use of time and
what a benefit to the public." Stossel's argument was notable for its
rarity in public. While many other academics more or less agreed that
regulation had gone to excess, public assertion of that opinion had been
rendered politically incorrect by the recurring negative official reports

on academic ethics, the related misdeeds of Big Pharma, and journalistic and book-length denunciations of academic-business dealings.

In response to Stossel, Korn reminded the audience that Congress initially showed a great deal of trust in the integrity of government-financed academic researchers. The rules, he said, were "remarkably light-handed for federal regulations, remarkably light-handed," he repeated.

> And they show a remarkable deference to universities and other institutions that received federal funds. They trust the universities and the universities' faculty to behave well. And they trust them so much that all you got to do if you're a grant recipient is provide what's called an assurance to your federal agency that funds you that Stanford University [for example] has adopted policies on research misconduct, on financial conflict of interest, and will implement that. "We have policies and will implement them." And, yes, there are some reporting requirements—that you're supposed to give NIH when an applicant for a grant has a conflict . . . and the institution has managed it. But you don't have to really tell anybody how you managed it. That's entirely the business of the university. And nobody looks over their shoulder.

Korn noted that his own research career began in 1961, "and it really was a lot of fun, and we didn't have all these regulations. We used to pour radioactive waste down the drain into the water supply. But the fact is," he emphasized, "these regs were driven by *bad* behavior, not by malevolent, powerful people who wanted to screw us up and tie us in knots." Returning to the argument about protecting science's financial relationship with government, Korn added, "We need the trust of the public and the Congress that is the source of most of our money. You look at the sponsored research in all American medical schools last year, less than 10 percent of that money is industry. Even though that percentage has been coming up from 2 or 3 percent, and it's okay, it's good that it's coming up, but 90 percent or so is still public tax revenue. That's a place where we've got to behave ourselves." Korn continued: "There were some pretty rotten reports that came out about the IRB system in jeopardy . . . rampant conflicts of interest among IRB members, rampant conflicts of interest among their institutions who were supposed to kind of oversee their behavior, and so on and so on and so forth."

Overreaction?

Then Korn made an admission rarely, if ever, heard in public from a leader of the medical establishment: "Was it hysterical and overstated? Absolutely, I believe absolutely. I believe absolutely," Korn repeated, "because I think a lot of the problems weren't anywhere as near as bad as that. But it doesn't matter. Nobody asked Dave Korn whether he agreed with the IG [inspector general]. Nobody in Congress called me up and said, 'Should we pay any attention to the IG report?' It just doesn't work that way. So Jordy Cohen, my president [of the AAMC], announced in his presidential speech in the fall of 2000 at the annual meeting that we were going to launch a task force on financial conflicts of interest that I had the pleasure of being responsible for.

"We don't endorse the hysteria," Korn continued, "and there is a huge amount of hysteria. In fact, I told a number of people that I think some of the British [scientific and medical] journals are getting a high on from bashing things American these days and are really going overboard in some of their hysterias. I'm not supportive of that," he asserted. "We think the relationship between academic medicine and industry is not only important; it's essential."

Korn concluded by endorsing part of Stossel's argument: "I fully agree with Tom that there are people out there, plenty of people in our community, who have given us no good by these flame-thrower, take-no-prisoner approaches to these issues. We certainly don't endorse them or agree with them. But I don't think we can go back to this Eden where everybody trusted that everybody was going to behave virtuously and a gentleman's word was his bond, and you don't need any kind of framework of oversight."

Rhetoric and Reality

Did the recommendations of the university and medical school associations beneficially affect behavior? Two years after issuing its 2001 report and recommendations on individual conflicts of interest, the AAMC surveyed its membership, now totaling 126 schools, to assess their responses. The survey revealed some progress, but no mass movement toward the ethical high ground. Fifty-two percent of the schools said they had increased "the protection of human subjects in research in which there were individual financial interests." Sixty-one percent used the rebuttable presumption or a similar standard against participation in human experimentation by researchers with a "significant financial interest" in the project. Sixty percent required researchers to disclose relevant financial interests when making oral presentations. And 74 percent required disclosure of financial interests on informed-consent forms. Seventy-six percent had established conflict-of-interest committees, but only 21 percent followed the task force's recommendation of inclusion of members from outside the institution. Finally, 22 of the AAMC's 126 member schools did not respond to the survey, leaving uncertainty as to what, if anything, they had done in response to the recommendations for loftier ethical standards. The text accompanying the survey report stated, "Although the findings are encouraging, they nonetheless indicate that more work remains to be done." [37]

Koski Looks Back

Koski's Washington experience might have soured him on the prospects for fixing the long-standing ethical problems in experiments with humans, especially under the George W. Bush administration. But when I spoke to him three years after his return to Harvard, he was surprisingly optimistic. Having served a frustrating stint in Washington, he doubted that government would lead the way to reform. Rather, he believed that correctives would come from efforts within the research community to establish an accreditation system that would examine the patient-protection programs at individual universities, identify shortcomings, recommend corrective steps, and certify those that demonstrated adherence to high standards. The accreditation would be voluntary, but, Koski concluded, universities would find it worthwhile to gain that stamp of approval.

> My guess is if the GAO went out to do a follow-up study today, they would probably find there's greater consistency in the guidelines and their implementation, at least at the major medical centers and all across the country than when they did their previous study five years ago. But I still think clearly we have a long way to go, but I think it is continuing to change. I don't think it's going to change simply because the government says let's change it. The real responsibility for change here falls on the institutions making a commitment to doing it right and probably upon outside organizations and processes. I think those are the things that are going to make a difference, because, to sound cynical, I for one, as someone who has been in government [and] worked on these things, am not

yet convinced that the government, number one, really wants
to have an effective system, or, number two, is ever going to
make the resources available to the agencies that are supposed
to do this job to really make sure they're going to work as effectively as they need to. I think that many institutions have
come to realize that they're probably far better off to discover
their deficiencies and fix them, take responsibility for them
on their own, rather than have the government do it. And so
we're beginning to see pretty much all of the major academic
medical centers move toward achieving accreditation of their
human-subjects protection programs. Why? Well, it's pretty
common sense, if you can tell the world that your program
has been accredited to a set of standards that go above and
beyond the regulatory requirements. If you're back in OHRP
and you have limited resources, where are you going to direct your effort? To those that have been independently documented through a critical review process that they're above
and beyond the requirements, or at those that have not made
that commitment?[38]

Koski drew hope from the creation of an organization to promote
and verify strict standards for the protection of humans who volunteer
to participate in medical experiments: the Association for the Accreditation of Human Research Protection Programs (AAHRPP, pronounced
"ay-harp"), which opened in January 2002. Inspired, as usual, by fear
that erosion of public trust would undermine appropriations for the
NIH, the new association was backed by the big professional organizations of university-based research, led by the Association of American
Medical Colleges and including the Association of American Universities, the National Association of State Universities and Land-Grant
Colleges, and the Federation of American Societies for Experimental
Biology. AAHRPP's purpose is to instruct universities to meet high
standards for the protection of medical volunteers and to verify that
they are doing it right. The accreditation process calls for a regimen
of self-assessment, followed by on-site evaluation by a visiting team
of specialists, review of the team's findings by AAHRPP's Council on
Accreditation, and a reevaluation every three years to retain accreditation. For this, the research institutions pay on a sliding scale linked to
the number of research projects they're conducting. The application fee
ranges from $8,100 to $26,000, or more for unusually large research
portfolios; and the annual fees are from $4,000 to $11,000, or more.

Whether, in fact, the accreditation process upgrades the sorry performance that has long plagued human-research protections—and that led to the birth of AAHRPP—is difficult to assess, especially at this early stage. In September 2005 AAHRPP announced that twenty-four organizations had been accredited, including Johns Hopkins, Massachusetts General Hospital and Brigham and Women's Hospital, the University of Pittsburgh, the University of Iowa, the University of Minnesota, Washington University (St. Louis), and the Western Institutional Review Board, a commercial IRB. AAHRPP said an additional two hundred organizations are at one stage or another of the review process for accreditation and added that it "estimates that in the next two to four years all the major academic research organizations will have completed the accreditation process." [39]

The New OHRP

In February 2002, shortly after Koski's departure from OHRP, Bernard Schwetz, a veterinarian and toxicologist at the FDA, was named acting director of the agency. It wasn't until April 2004 that he received a full-fledged appointment as director—a not-unusual lag in the anti-government Bush administration, which has consigned many senior federal posts to the uncertain status of acting appointments. While the NIH budget had doubled to $28 billion by the time of Schwetz's full appointment, OHRP remained a diminutive agency, with an annual budget of $7.5 million and a staff of forty. In a conversation with me in the early days of his tenure, Schwetz sketched a conciliatory operational strategy for OHRP, with heavy reliance on educational outreach activities to instruct universities in adherence to federal regulations for patient protection. As for attention to failings that might endanger volunteers in experiments, Schwetz explained:

> We are not out there checking every investigator, every study.
> We don't have the resources to do that. We depend on people
> who raise complaints to us that are the basis for things that we
> need to follow up on. Issues arise every day, from the stand-
> point of phone calls that we get, where people are asking ge-
> neric questions, or somebody will call and complain that this
> is something that they've seen, and they don't quite know if it's
> right or wrong. So what we ask is put it in writing and send us
> a clear description of what the problem is. Well, there are a lot
> of people who may decide it isn't worth it, and we never see a

letter. There are others who clearly write a letter that allows us to follow up. If it's a university setting, we get back to the university and tell them that we've had this complaint and we want to know, if in fact you are aware of it, is it real, have you done something to solve this problem, and if not, are you going to do something? Quite often they will say, yes, we're aware of this, and here's what we've done to solve the problem so it wouldn't happen again. In which case, we say, that's fine, and there's no further concern on our part.

Schwetz also expressed concern that some universities are overcautious and are taking on needless administrative costs and burdens in following federal regulations, fearing that "when OHRP stands ready to shut down the university, what do you expect us to do but be conservative?" As a result, he said, they "have put more into place than was absolutely necessary."[40]

In 2004 a descendant of the ill-fated "Draft Interim Guidance" that Koski issued four years earlier emerged in final form from the Department of Health and Human Services. Holding the status of official policy, it advised research institutions to identify individual and institutional conflicts of interest, evaluate their potential for affecting clinical trials, and determine the need for remedial steps. The new and final version was deferential to institutional autonomy. Absent was Koski's hard-edged language concerning the risks to patient safety and scientific integrity posed by institutions with a financial stake in clinical trials. The final version stated: "This document is non-binding and does not change any existing regulations or requirements, and does not impose any new requirements."[41]

: : :

Having looked at the setting and the system of science for sale in part 1, I will now examine personal experiences at the interface of academic science and commerce. Part 2 has lengthy conversations that I conducted with participants deeply involved in academic-industrial relations. Their interactions, or collisions, with commercialization are personally unique in some respects but also familiar in the contemporary scientific enterprise. Their accounts reveal a great deal about the innards of our subject.

**Part Two: As Seen from
the Inside—Six Conversations**

8 Success and Remorse

The movement of scientific knowledge from an academic laboratory to commercial success is rarely sure, smooth, or predictable. Even when, as infrequently happens, the process culminates in a marketplace blockbuster—the Olympic gold of tech transfer—friction between the parties sometimes develops along the route and progresses to conflict and even to litigation. Robert Holton, whom we met briefly in chapter 3, played a key role in the 1980s in the development of Taxol, a widely used and famously profitable drug for the treatment of breast and ovarian cancer. By 1991 Taxol, generically known as paclitaxel, was "the best-selling drug in cancer history," according to the General Accounting Office.[1] The GAO, tellingly, got into the picture when members of Congress charged that the manufacturer of the drug, the pharmaceutical firm Bristol-Myers Squibb, unjustly collected a bonanza from a discovery financed by the U.S. government. As with many highly profitable drugs, Taxol and contention are closely coupled.

Holton is a legend on his home campus, Florida State University, for his scientific and financial accomplishments. He is the source of over $200 million in royalties for the university, as well as considerable wealth for himself, all derived from his pioneering contributions to the development of the drug. Yet, as he tells it, the story of Taxol and commerce is not a happy one for him personally, nor, as he sees it, does it suggest wise policy making in behalf of scientific progress or financially prudent commercial dealing in drug development by the U.S. government.

The Taxol story began long before Holton's involvement with research that led to the drug, dating back to a government program in the 1960s that sought anti-tumor properties in plants and other natural products. Congress aggressively pushed the program, though many leading scientists argued that the money would be better spent on basic cell research. I recall the late Donald Fredrickson, then an institute director at the NIH, long ago telling me—with a skeptical grimace—that the cancer drug-screening program was predicated on the desperate hope that a healing miracle might lie undetected in the mysterious chemistry of some faraway patch of mud. The worldwide search and screening were extremely costly but were not based on unreasonable hopes; nor were they unproductive—just nearly so. Drugs derived from plants and other natural sources date back to the dawn of medicine and remain important to this day. Well known among medicinal plants, quinine, for the treatment of malaria, is extracted from the bark of the cinchona tree, originally found in South America and cultivated in Asia and elsewhere. Digitalis, one of the most widely used treatments for heart disease, is derived from the leaves of a common variety of the digitalis plant.

In the search for anti-cancer drugs, nature provided grudging cooperation. Between 1960 and 1981, the cancer program screened nearly 115,000 plants and over 16,000 extracts from insects, marine life, and other creatures, with virtually no success. One of the few promising prospects was found in extracts from the bark of the Pacific yew tree, which showed powerful anti-tumor effects in laboratory tests. As several scientists, with support from the National Cancer Institute (NCI), investigated the therapeutic potential of the discovery, they were confronted by yet another difficulty: the Pacific yew bark contained only minuscule quantities of the active ingredient, an extraordinarily complex molecule that challenged the science and art of chemical analysis and synthesis. If dependent on the bark of the slow-growing Pacific yew, an expansion to large-scale drug manufacturing would necessitate an environmental catastrophe—the massacre of the yew groves of the Pacific Northwest and, ultimately, destruction of what then seemed to be the sole natural supply. Thirty tons of bark from the Pacific yew tree might yield 100 grams of purified anti-carcinogen, at a cost of $250,000 a pound. Synthesis of the essential molecule became the goal of re-

searchers in the United States and other countries, among them Robert
Holton, an organic chemist who was fascinated by the challenging complexity of the yew-derived molecule that came to be known as Taxol. In
1985, after holding posts at several other universities, Holton returned
to his alma mater, Florida State University, as a professor of chemistry.

A High Priority for Taxol

Holton regarded himself as a fundamental scientist, focused on understanding the intricacies of complex molecules. His passion was science
far "upstream" from pills and practicality. But enthusiastically backed
by Congress, which pledged unlimited spending, the war on cancer
was rapidly expanding. At least partially to his later regret, Holton was
drawn into research narrowly targeted on turning Taxol into a practical, plentiful drug. Important as it was, the Taxol project was applied
research, lacking the prestige, and mystique, that the scientific culture
attaches to basic research, the quest for fundamental knowledge—the
kind of research almost exclusively blessed by the Nobel prizes in
science.

Politicians and the public bow, if with little comprehension, to the
importance of basic research. But they want practical results from
government spending on science. Taxol looked promising for curative
purposes. In the heavily funded federal cancer program, it held a high
priority. The race to synthesize a miracle drug produced ample support
for researchers, including over $1 million a year for Holton's labora-
tory group. By 1989 Holton was in the lead with a semi-synthetic process that utilized ingredients from a more plentiful plant source, the
English yew, an abundant European shrub. But tiring of the costs in
the seemingly endless pursuit of a plentiful bedside version of Taxol,
the NCI sought a pharmaceutical firm to pick up the financial burden. This turnover tactic was frequently used by budget-constrained
government agencies after early stage research displayed promising
therapeutic and commercial possibilities. In 1990 Bristol-Myers (later
Bristol-Myers Squibb, or BMS) licensed the Taxol-related patents held
by Florida State University and entered into a drug-development deal
with the NIH known as a cooperative research and development agreement, or CRADA, under which the NIH provided BMS with research
funds. And BMS provided money for Holton to continue with his research on synthesizing the difficult Taxol molecule.

Billions of dollars in research funding and prescription income eventually circulated through and around the tripartite dealings involving

the NIH, Bristol-Myers Squibb, and FSU, home base of Holton's laboratory. Such linkages were not unusual in the political atmosphere that encouraged tech transfer from academe and government to industry. And as the money grew, so did controversy over who did what to achieve success and whether the taxpayers, including the cancer-stricken among them, were entitled to "fair pricing" of a lifesaving drug that they helped to finance. BMS claims to have spent a billion of its own dollars to get the drug to market. Worldwide sales of the drug totaled $9 billion from 1993 through 2002, according to the GAO. Taxol, initially priced at over $5,000 per treatment, cost Medicare—the government's health-care program for the elderly—$667 million for patient treatments from 1994 to 1999. The NIH reported that it spent $484 million from 1977 to 2002 to develop, refine, and test Taxol. In return, it received a mere $35 million in royalty payments from BMS's sales of the drug. BMS agreed to provide Holton with $1.7 million to continue his research. By the end of the 1990s, FSU's royalties from sales of the drug exceeded $200 million.

As prescribed by the Bayh-Dole Act, Holton, as inventor, was entitled to a share. Under the formula adopted by FSU, his cut amounted to 40 percent, enabling Holton to pursue further Taxol research as he desired, without the recurring burden of persuading outside scientific judges, in government or industry, to accept his research plans and priorities. Holton thus became a rarity in science: a university-based scientist with plenty of money of his own to use as he pleased for research. The money underpinned Holton's stubbornly independent attitude toward the proper role of academic scientists.

"They Want Everything"

I met with Holton in a small conference room at Florida State University in February 2004 to discuss the Taxol story and his experiences as a career-long academic who became heavily involved with industry. Feisty, friendly, a bit grumpy, but straight talking, he took me on a verbal tour of the underside of tech transfer as he had experienced it over two decades—a tale quite different from the upbeat scenarios of tech-transfer enthusiasts.

Though I had read up on his long-running conflicts with his industrial partners, I began our talk with a bumbling attempt at humor, asking, "What's the secret of having happy relations between a university researcher and industry?"

"Well, I'm not sure I have happy relations," he stiffly replied. "I had a relation with Bristol-Myers, and that one always was pretty rocky. What was rocky about it was that it's the usual one-way street. They want *everything* and they don't give anything back. And that's the standard in terms of a research collaboration with industry. In the final analysis, it wasn't that bad a deal, because I learned to simply operate pretty much independently of them. But in terms of a research collaboration, I don't think I would necessarily recommend that."

"Were the difficulties peculiar to this particular relationship, or did you regard them as systemic in academic-industrial research relationships—a problem of 'different beasts' getting along?" I asked.

"I think it may be systemic," Holton replied. "They're very different beasts. I think the industrial side is always quite protective of their results and unwilling to share it back with the academic side. There was some stuff that wasn't completely on the up-and-up. Here's an example: Once we had the research collaboration established contractually, I went to Wallingford [in Connecticut, site of BMS's research center] for our first meeting. And we sat around, and the Bristol guys were suggesting all sorts of things that our part of the team might do. Well, some years later there was a dispute over who owned what. And we had made some compounds and the Bristol attorneys were claiming that the Bristol people should have been listed as inventors on inventions of *ours* on patent applications. And lo and behold they ponied up notes from this first meeting. And it was concepts that the Bristol guys had written down when they were attempting to tell us what to do. And this apparently was all premeditated. As they saw it, we have a meeting up front, we talk about what we're going to do, and we keep our concept sheet, okay, so later on, if that actually comes to pass, we can claim to have invented it."

"So, this was with malice aforethought. It wasn't just an honest misunderstanding?" I asked.

"Apparently," Holton replied. "They kept notes of this for that reason. It turns out that the lawyers couldn't interpret the notes, and they were way off base. But nonetheless, the intent was there. I think this is the way industry does business a lot of the time. Things are very different now. My Bristol days went from 1990 through 1995. And ultimately, that group at Bristol dissolved and went away. And Bristol ultimately turned all of that intellectual property back over to us. And since that time, I've founded my own company. Of course, that's a different story, a totally different story, and in terms of the collaboration there, it works fine. I've had experience with dealings between companies. For example, my company, Taxolog, has a collaboration with Wyeth [Pharmaceuticals]. And, I have to tell you, that's been just as unsatisfactory as my collaboration with Bristol. And when you get a collaboration between a small company and a big company, the big company always seems to feel that they can take advantage of the small company: 'We don't share our results, we make our decisions, we go our own way'—same attitude."

"The bottom line is what really motivates their actions?" I asked.

"That's been my experience," Holton said. "I remember when I was an assistant professor at Purdue, there had been a history of the Purdue faculty having industrial contracts. And as the Big Ten schools went, chemistry at Purdue was looked upon as something of a second-tier organization. And I heard the words 'handmaiden of the industry,' 'handmaiden of the industry' talked about quite a bit in terms of academic-industry collaboration. I think I'm basically against it. I think it just doesn't work, because the purposes are so different."

I remarked that "the front office at Florida State, as well as practically every other major university in the country, if not the world, is pushing for industrial collaboration. You've got these technology-transfer people asking, 'You got anything here in your laboratory that we might patent and license?'"

"I see those being different," Holton said. "If you can simply license a piece of technology and throw it over the wall, and somebody goes off and develops it, that can turn into a profitable thing for both parties. On the other hand, if you're going to do *collaborative* research with the industry, that's where you run into problems. I think we as academics should forget about the profit motive and go off and do what our curiosity inspires us to do. Now, in terms of Taxol, I was doing just that, and I just happened to be at the right place at the right time, where there was a huge problem, and I was able to provide a solution to it. But that's just because I was already working in the field for many, many years, and little did I know that this opportunity was going to come along. And when that happens, and you have something that works for them, and you can simply hand it off, that works."

"Do you think the people in the industrial labs are pretty much under the thumb of their front office?" I asked. "They're scientists, too, the people in the labs," I said, "and they probably share some of your objectives of good science and helping humanity, and so forth."

"In my experience, that has not been the driving force," Holton said. "The driving force has always seemed to be to get ahead in their own organization. They want to climb their ladder. The people in the front office are motivated by the bottom line. They need a product. They don't care so much whether the experiment was done well or not well. They want a product. And that certainly inhibits the people down the line from being very curiosity driven. That's a real luxury in the industry that I've been exposed to."

"You're portraying the industry as being rather narrow-minded and selfish," I said.

"Oh, yeah, I think so. I think that's fair to say," he assured me.

I noted that "Big Pharma describes itself as working in behalf of humanity."

"Of course. That's the PR department," Holton countered. "In terms of Big Pharma, I don't think there's very much truth to the altruistic motive. Not much, and certainly what I've seen of Big Pharma most recently, it's kind of, when you get down in the trenches, on a particular project there may be a dozen different departments. The people in those departments most of the time have never met one another. And the rewards, interestingly, for them frequently go to the department that has the most trouble. It's kind of like it's beneficial to have a hard time doing your job and holding up the development of the project, because next year, when the resources are allocated, if you were the weak link, the slow part, then obviously you need more money. That's a great system. And what you see in the trenches in Big Pharma is kind of every man for himself."

"You are describing a rather intellectually corrupt and irresponsible system," I said.

"Yes, I am. Unfortunately, I can't help it. I could help it in my own company. I can work for teamwork and rewarding those people who put in extra effort and go the extra mile and get the super results, and stuff like that. But I don't see that happening in Big Pharma."

"Big Pharma says it gives away pills to the poor. That's altruism, isn't it?" I asked.

"Yes, that's what they say," Holton replied. "That's the PR department talking. I've also seen the other side of the coin, where they've refused to try to develop a drug because the market is too small."

"Do you feel that Bristol made any valuable scientific contribution to the project?" I asked.

"They did a great job of getting it to market," Holton acknowledged, "but not in terms of anything that would be a scientific advance. The first thing they did was they went out and found some resources and set up a network for harvesting bark and extracting it. Okay. The next thing they did was they scaled up our synthesis process and they did that very well. And they transferred that to a plant they newly built in Ireland. And they did that very well. And they did it pretty much in record time. And they got the drug approved and on the market, and they did that very well also. But I wouldn't say there were any scientific advances per se. They really didn't do anything new on our semi-synthetic pathway. They pretty much took what we had done. It scaled up fine. It worked. Of course, they had a research collaboration, the research team in Wallingford. But they really contributed very little."

"What were they doing?"

"They were doing *stuff,*" Holton said derisively. "Okay. But you have to understand, there was an awful lot to be done. And what you're trying to learn in an activity like that, you're trying to learn chemical structure-activity relationships, and you're looking for new ways to reshape the molecule to make a better drug. Well, that means you have to have a certain efficiency and a certain throughput of new compounds and biological data coming back in to try to make that picture. In Wallingford they had a group of very talented people, very talented chemists, and it was like they were in a new world, because you have to understand, typically in the pharmaceutical industry, chemists don't work on very complicated compounds. I call it 'the world of flat molecules.' Or I might put it 'the prison of flat molecules.'"

"Why don't they work on complicated compounds?"

"Because it takes too long to get a product out the door. It's too complicated, too intricate, too much stuff going on. They want things where they can produce a result in eight months to a year. Maybe even shorter than that. You can take a big old complicated molecule that's got all the stuff on it—it takes you a long time. Now, the interesting thing about Taxol was all these bright young guys out of Harvard and Yale and MIT and whatnot that were in this group may have been in the 'prison of flat molecules,' and all of a sudden they got to work on Taxol, where you throw in a reagent, and it does crazy things. It rearranges this way and it rearranges that way, and so forth, and they were just having a great old time trying to figure out what the molecule had done. And while they were having a great time, it really didn't lead in a productive direction. So actually, by the time we'd been in that collaboration for a couple of years, I had to start up another group here to do what they should have been doing in Wallingford. Pretty interesting. No, they didn't get much done. Which is, interestingly, why we wound up with all the patents."

"In a professional, comradely fashion, did you have occasion to talk to some of their people and ask them what the hell is going on?"

"Sure," Holton said.

"Were they ashamed of themselves?"

"No, no," he explained. "They were just interested—you know, they did some good chemistry, but they would go, like I said, go find a molecule rearranged seven different ways and be spending their time figuring out what it did. It led to some very interesting stuff, but it wasn't making new compounds."

"So you got the patents when you did the work?"

"Right," he replied. "And, interestingly, well, I was very fortunate to have a great patent attorney and have a great relationship with him."

"Was that through the university?"

"Well, yes, it was through the university, but this guy works for a firm in St. Louis that the university retained, and after our first couple of patents, it was pretty much that I talked to the patent attorney, and we bypassed the tech-transfer office and all this stuff, and we were under a deal with Bristol where Bristol was paying for the patents, and so it would go like this: Something happens in the lab here. I call up the patent attorney. I send him a page or two. He generates a patent application. He sends a copy simultaneously to the patent office and to Bristol, and the bill to Bristol. It was pretty simple. You have to realize, though, that at the end of the research period, in 1995, Bristol took a look and said, 'Okay, we'll keep a license to the synthesis of Taxol. But all these other patents and all these other compounds and stuff, our research group got mad and quit and they said there wasn't anything there and, oh, by the way, we don't want to keep this. So, we'll just stop paying for these patents. FSU, you can have it.'"

Keep Industry Out of Academe

Turning to another topic, I asked Holton, "Do you see any problems in academic researchers accepting money from industry to conduct research?"

"Yeah," he replied, in a tone that suggested that the question was inane. "Why is the industry paying for the academic research?"

"The academic may have particular skills, or industry doesn't have the facilities or the people," I suggested.

"Well," Holton said, "that basically makes the university a contract research organization. There are contract research organizations out there that provide all kinds of services. I don't think that was what we were here to do. I thought we were here to do basic new science—curiosity driven. That's different from providing a service."

"You've been in the life sciences for a long time, in chemistry," I said, "and you really have very serious doubts about doing work on contract for industry?"

"Absolutely. I'm against it," Holton said emphatically.

"Even if the work is of a basic nature, the same as what the NIH would support in a university?"

"I'm not aware of any time when that's happened," he replied. "There might be a way that you haul it up and twist it around so that could happen, but I've never seen it happen. I used to have a friend who used to be at some school down the road, and he would support his research by getting contracts from various people in the industry to make compounds for them. 'I want X amount of Y compound. Can you make it for me?' 'Yes, I can.' 'What's the price?' Negotiate the price, so forth. And operate as a contract synthesis laboratory, bringing the money in through contracts and grants and basically use it to pay his grad students and postdocs. And his grad students and postdocs would spend a portion of their time working on these contract syntheses."

Holton paused, slowed his delivery, and, with emphasis on each word, declared, "I don't think that's why we're here, and I hope we never become so impoverished that that becomes the mold."

"In an ideal world, maybe that's not why we're here," I pressed on. "But on the other hand, there is practical necessity. Universities have financial constraints. It's a way to make some money that doesn't fundamentally undermine what the university is doing. Universities make money in lots of different ways. If you have laboratory facilities and the people, what's wrong with them taking some small percentage of their time and bringing in revenue?"

"What I see wrong with that: It's *not* in our mission," he said. "We're here to try to come up with new stuff, new ideas, and stimulate the people who work with us to think in new and different ways, master the techniques that we have, and become innovators in their field. And going out and painting cars part-time to make some money to pay for this simply diverts those energies. That's why we have things like the NIH, the NSF, and presumably, theoretically, if you write down an idea that people think is interesting, then you can get money from those agencies and proceed in that way."

"That's a very idealized portrait that you paint," I responded. "But historically, American universities have performed a great variety of functions, from building the atom bomb to telling farmers how to keep worms out of their peaches. These have been far from fundamental research."

"Probably," he reluctantly conceded.

"You are painting these other activities as being unclean and inappropriate," I said.

"No, I'm not," Holton replied. "No. I speak in the context of my experience. I am a synthetic organic chemist. I am a chemistry faculty member. And I'm speaking to you as that. I'm not in the Agricultural Extension Service. I don't know how they do business. They can do business any way they want to. That's my purview. I speak from my experience."

"As far as your sector of the university is concerned, you want to stick to fundamental research."

"Absolutely," he said. "It doesn't make any sense for me to be sitting here having a bunch of people distill solvents to ship out and sell to the industry. It doesn't make any sense at all."

"Intellectually it doesn't make sense?"

"That's right," Holton declared. "This is an intellectual endeavor. And we need to try to keep it that way."

"How do you think we're faring in keeping it that way?"

"Could be better," he said, adding, "That's a question, though, that begs comparisons. Compared to what? How well could we be doing? There is more money going into basic research at NIH and NSF. That's good. There probably ought to be a lot more than is going in. How are we faring with that? I won't say we're doing real well; I won't say we're doing real badly. I think we're just kind of getting by."

I asked, "What sort of impression do you think students get when they see what's going on in university science departments? You do have people who are cooking up solvents for industrial customers. Is that kind of turning their heads away from where they should be?"

"I think so. I don't think it provides a good image. I don't think it provides a good role model. I don't think it's the right place to be. I think we ought to be in purely intellectual endeavors."

Get Down from Your Ivory Tower

"If industry wants to connect to university research—apart from reading the journals—is there any feet-on-the-ground way that you think they can do it without contaminating the values here?"

"That's tough. It's pretty hard," he conceded. "As a very general question, I'm sure it's specific to each case. The motives of industrial organizations are not necessarily the motives of the individual people in a university. There are ways in which it could happen. If you take my own example. I'm sitting over here doing purely esoteric NIH-funded stuff, trying to figure out the total synthesis of a big molecule, Taxol. One day somebody calls me up. It happened to be my program director at the National Cancer Institute, who also happened to be the prime proponent of Taxol, and he was on top of everything that was happening and says, 'Look, this is going to be a drug and we need a way to make it. So you need to bring yourself right down out of your ivory tower and come up with something that's practical, buddy.' Now, it didn't have to be the guy from NCI who did that. It could have been somebody from industry. If you pick somebody who's worked in a particular field for many, many years and has done fairly well and you have a problem which is right on point in their area and you contact them and say, 'Look, can you do this?' That might be a way it could work."

"But," I suggested, "you would have grave misgivings about giving industry any influence over what's going on inside academic laboratories."

"I certainly would," he agreed. "I even told the guys at Bristol in our first meeting, 'I'm going to do whatever I damn well please; you're not going to have a thing to say about what I do. Might as well know who we are right up front.' There wasn't much they could say. They could keep on making suggestions, but I wasn't necessarily going to do any of it."

"You knew how to make the molecule," I said.

"Yeah," he agreed—with satisfaction.

"You were in a very, very powerful position. But not many people are in that position."

"I think that's one of the reasons why boneheaded people like me go to universities," Holton said.

"A university is a refuge of the boneheaded?"

"A refuge of the boneheaded," Holton chuckled. "There you go. Independent, boneheaded, stubborn, mule-headed. You can call it anything you want to."

"And the basic research funding agencies in Washington stake them. Right?"

"Stake them. That's right," Holton said. "From my experience, I think the world's a better place if academicians go to whatever in the world they think they're curious about—whatever they think the cutting edge might be. And industry has a different motive, and it's those points of intersection where the academician can feel that he's not diverting his mission to do theirs. There those things might work."

"Where does the translational activity take place?" I asked. "You can't simply take what's happened in an academic basic research laboratory and move it into a manufacturing plant. There does have to be that bridging activity that makes it—called applied science, or what have you—that takes it from academe to the marketplace. But if it's a dumb-headed company like some you've been describing, they won't really get very far."

"That's correct," Holton agreed. "They're going to need a lot of help. That has to be from the developer of the technology. Right. If you think it's important enough. You've always got to trade off what you're giving up, because there's no free lunch. There are twenty-four hours in every day. What you spend during this hour is a choice. What you do during this hour is you're neglecting a whole host of other things."

Since Holton had been speaking mainly in response to my questions, I asked him whether there was anything he'd like to add to our conversation. "Well, as you can see," he said, "I'm not a big fan of industry-academic collaboration. I just happened to be in that spot where I was the right guy at the right time. The problem was right in an area I was working on. And I said to myself, 'I said I'd never do stuff like this. Well, okay, we got this thing—it's going to become a drug; they have no way to make it; they're sawing down half the trees in the world. Sooner or later in this total synthesis, I'm going to have to get to that part, so at the end maybe I ought to divert a little bit of resources and look at that problem now.' It was one graduate student for six months. Maybe it was eight months or it was a year. I don't know, but it wasn't more than a year. But I thought that's as close as I'm ever going to get, because I thought about that. Now, I'll tell you this: It has cost me *dearly*."

"Cost you dearly in what currency?" I asked.

"In academic currency," he replied with a clear tone of regret. "It has cost me dearly, because what I like to do is go out there and take on the synthesis of really complicated molecules and try to do it in a new and novel way, something that pushes the envelope. Well, I was doing that at that time. And while NIH has provided resources to my lab for years and years and years and years and years, there was never enough. And we're always living in poverty, in a relative sense. And in the years when Bristol came along, I was in one of those. It was a big 'war.' It was the biggest molecule that had ever come around in our field. There was a big war to see who could get there first, and the best. I like best better than first.

"Well, I needed resources to do that. And I had a lot of money from NIH, by NIH standards. But relative to what was really needed to get that moving, it was marginal. So the Bristol collaboration offered that—an infusion of money. It was a huge amount of money we got from Bristol, relative to the amounts that we had gotten from NIH in the past. Relative to an NIH grant, my ship had come in. And I was able to use those resources, although the Bristol group in Wallingford was not a resource, I thought it would be. But it wasn't. But I did use their money, and we did win that race ultimately. But we also made analogues [variations of the Taxol molecule], and we patented a lot of analogues. And in 1995 what happened was Bristol had gone away. We had patented—just a rough guess—somewhere in the neighborhood of 10 billion Taxol analogues. There is the possibility, if you can find it, that there is a lifesaving drug in that package. And I've tied it up so that nobody else can do anything with it, and nobody else knows anything about it. I'm stuck. I have to go try to exploit that. I would rather be over here in the lab spending all of my time thinking about the struggle and how to synthesize the next big molecule. But no, I've diverted a huge amount of effort and money, and so forth into seeing if we can't find the lifesaving drug in that portfolio. And that's what I mean by being trapped. That's not bad. I'll take it. But if I had my druthers fifteen years ago, I think I would rather have spent all that time in the lab just thinking about making ever-more-complicated molecules and better ways to do it."

Holton hesitated for a moment, and then said, "I'm in the middle of the struggle, man. Ask me when the ninth inning rolls around. When the ball game is over. But you see what I mean when I say I got trapped. And from the standpoint—well, you know. If I had my druthers, I would rather have not probably been trapped. I'd rather just— I picked what I wanted to do from the get-go. But it's not bad, and I hope we can do some good. We'll try. But people who work in my group here don't get as much out of me as they should, as they could. Let's put it like that. I can't be totally devoted to thinking about the projects that they're doing. Obviously, I take some time and I'd rather not. There's no option."

9 A Congenial Partnership

Robert Holton's sour experiences with industry contrast sharply with those described to me by a young researcher at Georgia Tech, Robert M. Dickson, an associate professor in the School of Chemistry and Biochemistry. Modest in manner, and frankly acknowledging his unfamiliarity with the intricacies of tech transfer, Dickson easily chatted with me on January 25, 2005, in his campus office. The topic was his research and a relationship he was just starting with an industrial firm, Invitrogen, a billion-dollar California-based company with forty-five hundred employees and a strong presence on the frontiers of biomedical research, which is where Dickson has been successfully working. In collaboration with scientists from nearby Emory University, Dickson and his Georgia Tech team focused on utilizing atom-size clusters of gold and silver that can cling to cancer cells and help produce *in vivo* images of the cells at the molecular level. The potential of this research for medical diagnostic and other purposes attracted considerable interest and financial support from the National Institutes of Health, the National Science Foundation, and the private sector. As best as can be foretold amid the uncertainties of fundamental research, tech transfer, and the marketplace, the future is bright for a scientifically productive and commercially successful outcome. But, as became evident during our conversation, academic-corporate collaboration also raises sensitivities about the flow of scientific information with commercial potential and other issues that ensue from science for sale.

After publishing a paper about the research in a premier journal, *Physical Review Letters,* in August 2004, Dickson explained, he was contacted by several companies that had seen the paper or press releases that Georgia Tech had issued—customary horn blowing by universities eager to tell the world about their accomplishments. As required by Bayh-Dole and Georgia Tech procedures, Dickson had disclosed the publication to the tech-transfer office, which took steps to protect the intellectual property rights and patent potential for Georgia Tech. Dickson recalled "discussions with many companies, and I went out to visit several of them to give talks. I always consulted with tech transfer first to make sure that I'm not doing anything wrong. And then after nondisclosure agreements and all that kind of stuff, and showing them more technology, a licensing agreement was just signed, where five people from this biotech company came out. They had been working long, long hours wordsmithing with various lawyers on their side and then they hammered out the final agreement. It's an exclusive license for medical imaging and diagnostics and R&D reagents, and that sort of thing." The focus of the deal, he said, is the biotechnology market, "but there are other potential markets as well."

Under the agreement, Dickson said, the company will provide funds for his lab. He wouldn't say how much. "I'm sure they're happy to do that, because it's much cheaper to do research at a university than at a company," he explained. "However, the nice thing about that is all IP [intellectual property] still belongs to Georgia Tech. So, even though they're funding the lab, and they probably have right of first refusal, all of the IP still belongs to Georgia Tech, yet they would have the right to license it exclusively."

The corporate funding, he said, will enable his lab to "basically continue along the directions that we're continuing in anyway and in part to build a good relationship. But I'm sure they're hoping that the work that we generate will be along the lines that they're interested in, and therefore generate new results that will help them commercialize things later. From our point of view, we're more interested in understanding the fundamentals. They're more interested in knowing the fundamentals, but also in whatever they need to do in order to commercialize."

"Do you regard this as a happy development?" I asked.

"Oh, absolutely, absolutely," he replied. "I think it will be a great long-term relationship. People came out here from the company, and

I introduced them to several of my colleagues to sort of advertise the work of my colleagues, and they also seemed quite interested in some other people here. There may be opportunities for some of my junior colleagues, which would be fantastic." He noted that tech transfer is strongly supported by the administration at Georgia Tech. "Talk with Charlie Liotta [vice president for research and dean of graduate studies] or Gary Schuster [dean of the College of Sciences]," he suggested. "I remember Gary indicating that one of his visions of the future relates to faculty members developing technology and then having outside companies invest in it, and that being the model for the future success and funding of a university. There is a strong effort to encourage tech transfer and licensing of technologies as well as start-up companies here based on technologies that Georgia Tech has developed, or even nonfaculty people in the community. If someone has a good idea, they can help out with laboratory space and some resources."

Fundamental Science, Promising Applications

"Your work sounds quite fundamental. A long way from the market," I said.

"Yes, one would kind of think so," Dickson agreed. "I'm interested in the cluster physics of how you go from a single atom to sort of my gold ring of nanoparticles. Yes, it's very fundamental science, but that fundamental science, because of the fluorescence of gold and silver, actually has some pretty exciting applications. Certainly, there's a lot of work to be done, but we've already shown we can make these water soluble, and they have some incredible optical properties that would compete favorably with anything out there. As a result, what remains to be done is further characterize the properties, purify things better, get higher concentrations of things. But then figure out how to attach them to proteins, how to label things, and how to develop a new class of reagents. NIH and NSF are helping us with this, and we've formed a center with several of my colleagues through NIH funding in order to make the tools readily available to people. So, we'll be developing the fundamentals, but there are companies who are actually interested in taking those fundamentals, transferring them to the marketplace, and then making them generally available, which is the best-case scenario."

"When you say 'make the tools,' do you mean transfer materials to other interested researchers?"

"Absolutely."

I asked about arrangements for publishing or otherwise publicly circulating information about the research, including the involvement of graduate students and postdoctoral fellows in the research.

"For work that's funded from the company—again, I haven't seen the full document—there's probably a delay of at most thirty days, so that we can make sure that if there's anything that we've discovered that's of interest to the company, we can make sure to file it here. Again, Georgia Tech will retain the patent rights, unless it's done collaboratively, in which case it will be joint patents. So, they can basically say, 'Oh, this looks awesome; this needs to be filed.' Then we'll file something to protect our interests both here and abroad. But then we can publish stuff after that thirty-day period. And that's really the only restriction on anything. Initially, the company was concerned, 'Oh, we only want the postdocs working on this that are paid by the company, and we don't want any other people doing it. We don't want them working on anything else.' I was like, my lab is too big and too free in terms of information flow to have such a situation, and even though it would fund two postdocs and supply money, it's not worth it to me to have them off on their own. I have other resources. Right now I have eleven people—eight graduate students, two postdocs, and one assistant. And, hopefully, one or two more postdocs."

"So you're sufficiently well funded from a variety of sources that you don't have to be—"

"I will never say that I'm sufficiently well funded," Dickson replied. "Things have gone very well and I've been able to support a good-size group, because my students are working hard, and postdocs are working hard, producing a lot, getting results. And we're able to take that and translate that into funding for our efforts. I'd like to have a steady state of ten to twelve in the lab. Twelve is a reasonable number, especially if I can continue with the tech transfer and secure industrial funding for longer term. The relationship right now is a three-year thing, and if things go well, I'm sure they'll be interested in continuing that."

"Does the industrial funding in any way interfere with what I gather is your primary interest, basic research?"

"I don't believe it does," Dickson said.

"They don't try to get you 'downstream' to more developmental things?" I asked.

"No," he said, "in terms of any sort of statement or discussion of work, that's basically hammered out between me and a sort of director there at the company, who's also a scientist. He really knows that their job is more scale up, and they take the lead from what we need to do anyway from a fundamental point of view."

"Do you find that being involved in this tech transfer imposes any kinds of restraints on your discussions with colleagues at other institutions?"

"No. Possibly with the industrial funding, because they won't want me to talk about the newest results before they've had a chance to sort of look at them, but again it's a question of where the research funds came from in terms of what I can talk about. And so, I think it would be more of an accounting or organizational thing as to what things I can talk about and what things I cannot."

"What do you mean by an accounting sort of thing?" I asked.

"Well, if it's projects supported by the research funding from the company, supported by them, that's not something that I should talk about too early. I should make sure that it's okay with them, or I should send them a document, or we should have protected our rights because the company will be interested in those things."

"So to some extent then, you do have to sort of segment your work," I suggested. "There are activities that are funded by the company, and then there are various others supported by NSF, NIH, Georgia Tech itself."

"Yeah. I mean certainly you want in working to make them all orthogonal, perpendicular, so that there's no overlap. But because there are different aspects of these metal nano-clusters and how to make them for bio-labeling purposes, there is some overlap, but it is a synergistic overlap. So, we're funded by NSF and NIH in different complementary directions and funded by the company, also in a different and complementary direction. In fact, NSF and NIH really want the information to be out there for everyone and also the materials and methods to be out there for everyone. I've received a lot of requests actually to provide materials to people or to collaborate with people, and I can't do it. I mean—"

"Just simply the volume of it?" I asked.

"Yeah, I happily send off procedures—this is published information, how to do this, and try to help people out with that, but we're not set up to produce a lot of this stuff, otherwise that's all we're doing, and we're not able to do the fundamental studies that we want to do. So, it's actually much better for a company to take this information, scale it up, and be able to distribute it to people for a reasonable price, if that's what they have to do."

"Is the company doing that or will they?"

"They will, but they're not doing it right now. Again, because it just started and the licensing thing just went through."

Protecting Intellectual Property

"When you were coming up through graduate ranks and a postdoc," I asked, "did you foresee that you'd be or did you have any interest at that time in getting involved in tech transfer?"

"No," Dickson explained, "the reason being that what I did as a graduate student was so far removed from anything practical. The University of Chicago is also so far removed from anything practical that—or at least seemed to be when I was a student there—that it never really crossed my mind. When I was a postdoc, my postdoc adviser actually did have press releases about certain things, but not so much on the tech-transfer side. So, if there's a high-profile paper published in *Science* or *Nature* or *Physical Review Letters,* then she would be in contact with the press release office because they get the tables of contents and whatever and they wanted to advertise the work. That's something that they had done, and I was a little bit involved in that, at least about press releases. So, a little bit of experience in interacting with folks and generating a nice story that—and that's very useful for the university; it's very useful for the group. In terms of tech transfer, though, I hadn't really had any experience with that. When I came here, I guess I started thinking about it, learning from colleagues or learning that one should always protect their interests if there's any remote chance of a commercial application. Then just send stuff off to the tech-transfer office at the same time you send it out for publication."

"You mean PIs [principal investigators] and more seasoned people would tell you that's the way the world works today?"

"Right."

"Were they warning that something might be taken away by somebody who really wasn't legitimately entitled to it?"

"You know, people told me they've heard stories about that, or they're just protecting their own interests because they're a little bit more sort of savvy or aware in terms of the IP than I used to be and probably still more than I actually am. So, tech transfer here really helps me with that, and I rely on them heavily for these types of things. Honestly, I don't want to be involved in the licensing agreements."

"Have you ever personally encountered or heard of IP being misappropriated by some researcher who really isn't entitled to it—just filching it, essentially?"

"I don't know anything firsthand, or really even secondhand," Dickson said. "I have heard stories of folks who go out and talk—you know, academia is reasonably free flowing, where ideas are reasonably free flowing. And so, I have heard of people who go out and give talks, sort of what I do, about things that they've just discovered in the lab and just figured out how to explain and want to present the latest, greatest results. And if there are some people in the audience who know someone at a company or are involved in a company or—I don't know that it's malicious. But I have heard of people basically taking that information and starting up companies or patenting that or whatever. And somehow the PI being locked out. I honestly don't know how true those things are, but I'm sure that type of thing happens."

"Do you feel, then, some element of caution is warranted?"

"Yeah. Yes, without a doubt," Dickson replied. "You know, if you worked hard to develop things, you want to be protected. Again, I'm naive. I don't really think about the financial gain, I'm more thinking about will it enable me to do more research and that sort of thing. I'd hate for a research project to be taken away because a company basically determines, wow, this is like a hundred-million-dollar-a-year or a billion-dollar-a-year-type project. And even if we've protected ourselves, they could basically put fifty people on it, and then I wouldn't be able to compete with them. That would be a bad thing. It would be bad if someone else were to make money on ideas that we generated, and we were just foolish enough where we didn't have the foresight, or tech transfer didn't have the foresight, to actually push it forward into a patent. I guess that one of the downsides in academia is lack of money, lack of resources, such that, my understanding is that in companies every paper, every proposal certainly gets reviewed by a committee and they determine whether or not they should patent it. Here, we really don't have the resources to do that. It may pay off in the end to do that, but we don't have that setup right now. Tech transfer will only push forward through the patent application and all the time and expense involved if a company is interested in it. And that's got to be basically within one year of when we file it. So it's probably going to be at a stage where it's more or less ready to go."

"When you give a talk about what's going on in your lab, depending on the audience, do you feel that prudence calls for being a little bit cautious about what you tell about in fine detail?"

"The experiments in my lab are sufficiently difficult that I'm less concerned about that," Dickson explained. "In terms of the overall ideas, I'm not really all that concerned about that as well. If people need details for preparation or people get excited about it, they'll contact me and then we'll be a little reserved about the information we give out, depending on who it is. If it's from a company, we may be a little more concerned. If it's from an academic colleague, especially someone I know, sure, we'll tell him exactly how to do things. If one is sitting through one of my talks, one really has to be able to see the big picture and to figure out what the applications are. I say what some of the applications might be, but again I focus more on the fundamental physics with some special applications of it. And that should get people interested in it, but they're going to have to come to me to figure out how to get the stuff done."

"Why would you not put as much clarity into the talk about the applications as about the fundamental work?" I asked.

"Because when you go to a university and give a talk, or if you go to a specialized conference, then you tend to focus more on just sort of the beauty in the physics or the physical chemistry or the chemistry. And you can talk about some applications, but really what's exciting as far as I'm concerned is not the money you're going to make off of it, but more just how the fundamental chemistry or physics really enables this and what the scaling relationships are and what we've discovered. Okay, there are applications, but I'm not going to talk about that. I don't want to get into being someone who talks about the applications. I want to stick with the fundamental physics. Going back to when I was looking for jobs, or my academic history, there's an interesting time when I visited Bell Labs before it sort of was disbanded. But in October of '97, I was thinking about looking for jobs at that point and I went to visit a friend of mine there and he gave a talk, and I just remember being absolutely awed by the basic research, yet the practicality of it, and really wanting to be able to do something practical. I'm quite happy here that I've been able to still do the fundamental science, but it can have a more applied thrust. And that's one of the advantages of being at Georgia Tech, simply because you have the engineering influence, you have the engineers to talk to, and even the chemists have more of an application bent to their research, even though it's still very fundamental."

"Are there opportunities to share in the financial rewards, assuming some do come along down the pike?" I asked.

"Roughly speaking, any royalties, licensing royalties, any income to Georgia Tech gets split three ways, more or less equally, which is one-third to the inventors—which again my student is a co-inventor— one-third to the department, and one-third to the university."

"Has any money flowed so far?"

"For the licensing agreement, a check will be cut. There was a certain amount of money to secure that licensing agreement. The department doesn't quite get a third for the first—up to a certain level they only get like one-sixth, and Georgia Tech gets 50 percent. But the inventors actually get one-third of that. So, it's in our best interest and all of our best interests to continue developing things from a research point of view, potentially from consulting, potentially from lots of different things, in order to help move that technology along, so that they can commercialize it. That will bring more money into the university as well."

"When you're sitting in the audience at a talk by a colleague or someone who's involved in science or engineering that you're interested in, have you observed any kind of restraint on their part because of commercial considerations? Have you felt they're not telling the whole story?"

"You definitely get at times the opinion that people aren't telling the whole story," Dickson told me. "But I usually don't hang out with the people who have the more commercial interests. When I do go to a talk, sometimes you don't learn all the details, certainly, but I always assume that's just scientific things and whatever and you're sort of there to learn the big important concepts and some of the details. But you can always go to the person afterward and have a private conversation with them, and ask them how they did this and got wonderful results and that sort of thing. I would save those details for a more one-on-one conversation instead of a sort of big lecture hall asking a question relating to the finer details. Asking about clarification or asking about the applications or the interesting physics or chemistry, certainly. But details about how to do it or commercialization, most people would do afterward."

"I realize you can't see what use might come of this technology, but you must have some educated speculations about possibilities."

"Possibilities. Bio-labeling—one could imagine anything from security measures, like making sure something is genuine, to diagnostics for medicine. What we tend to push is more water-soluble fluorescent species, to replace organic dyes, not in clothing but in biology applications. And so right now there's a huge business for diagnostics for labeling, for looking, from a fundamental point of view, at protein interactions. So all of those are certainly possibilities. I've been amazed by some of the things that companies have come up with and suggested, things that I would not really have thought about. Not even just companies. There was one person someplace from a government agency interested in possibly using these things for fingerprint identification somehow. There are a lot of potentially interesting things that I hadn't thought about, that more people who are exposed to this and see this come to me with ideas, and they could license the technology for that purpose. It's really not my job to come up with those applications, because I don't have the expertise to come up with those applications. And I'm not someone who has a technology and is looking to shop it out to all these different places or trying to come up with things to develop that. I'm not starting up my own company or anything like that. What I'm interested in is concentrating on the fundamental physics and chemistry, and if that enables other companies or other people to get excited about things and go off with them, well, we're protected with the fundamental enabling technology."

"Could you foresee yourself going corporate industrial someday?"

"I would never say never," Dickson replied. "I really enjoy the academic environment, interacting with students, the freedom that I have. If such a situation could possibly be worked out where I could still have economic or intellectual freedom to go down certain paths, that could be attractive, but the academic lifestyle is just ideal for me. And many of these companies with fundamental, basic research are disappearing. Bell Labs is having serious problems. IBM still has some very basic research going on with an applied bent, which is great. I really enjoy my job, I really enjoy this place."

: : :

In November 2005, nearly a year after our conversation, I e-mailed Professor Dickson to inquire about further developments in his laboratory and in his relations with the company that had licensed his research. He responded that any information would have to come from the technology-transfer office. An inquiry there brought a response from the press office, to the effect that no further information would be provided.

10 When the Rules Change in Midstream

When we met on April 24, 2005, Professor William S. M. Wold was perplexed and feeling ill-treated by Saint Louis University, where he had worked for thirty years. The university, known as SLU, for short, was uneasy about the propriety of the complex financial arrangements for Wold's much-admired cancer research. The arrangements, linking Wold's academic research to a private company he owned, had been openly in place for several years without troubling anyone. But two changes had occurred during that time: sensitivities to conflict of interest had intensified throughout academic science, and SLU, aiming for a bigger role in scientific research, was concerned about its image. Wold was not accused or even suspected of wrongdoing, I was told by university officials. In fact, they emphasized that he is a prized member of the faculty. But appearances count, and so did the possibility of negative publicity for the university concerning his financial arrangements.

Wold is chairman of the Department of Molecular Microbiology and Immunology in the Health Sciences Center of the School of Medicine at the university. The second oldest Jesuit university in the United States, SLU was traditionally a blue-collar school and a latecomer to the growth ambitions prevalent today in higher education. Nearby Washington University is the major-league school in St. Louis, though both have similar enrollments, of eleven to twelve thousand students. In 2004 Wash U possessed an endowment of over $4 billion and spent over $475 million a year for research, mostly from government agencies, but also from industry, philanthropies, gifts, and its own resources. SLU's endowment was $700 million and the research budget was $55 million. But SLU was ambitious and stirring. Money for research was on the upswing, including over $1 million in government and industry awards to Professor Wold. As part of its growth program, SLU aimed to increase the financial yield from licensing discoveries in its laboratories—a mere $1.5 million in 2003. Toward that goal, SLU hired a new director and added staff to its technology-transfer office.

Wold was well-situated to help SLU achieve its ambitions. In addition to his academic position, he is president and CEO of a biotechnology company he founded, VirRx, to develop a promising cancer treatment derived from research that he and colleagues have conducted at the university. SLU is proud of Wold. But when I met with him in April 2005, he was under scrutiny by the university's newly established Conflict of Interest Committee. The chair of the committee, Professor Kathleen M. Farrell, explained to me that the rules governing outside dealings by faculty and conflicts of interest were in flux. SLU, she observed, is "a small, emerging research institution. We were not as well prepared to tackle these problems," she said, noting that SLU lacked the staff resources of major research universities.[1]

SLU's conflict-of-interest rules were first issued in 2003, several years after most other research universities had taken that step. A faculty committee was then appointed to review requirements for disclosure of outside income, consulting arrangements, adjudication of conflict-of-interest issues, and other details of ethical good behavior. Mindful that scientists are hypersensitive to bureaucratic intrusions on their work, SLU's administration proceeded diplomatically and cautiously to promote understanding and cultivate campuswide support for the new regulatory framework. Wold's exasperation was evident as he patiently related to me his experiences with what he regarded as arbitrary and unfathomable bureaucratic forces. The grounds for the university's concern about the complexity of his financial arrangements were also evident. The result was a collision between old and new ethical standards in the conduct of academic research.

Financial Complexity

For five years, openly, and with the approval of the SLU administration, Wold had straddled academe and commerce, with a patchwork of support from several U.S. government agencies and industry. He conducted basic research in his SLU laboratory on campus and drug-development research in his corporate laboratory. The corporate lab was a spin-off from his university lab, but in contrast to the typical academic spin-off, it hadn't spun very far. Legally, the two labs were separate entities. Geographically, they were close neighbors, with the corporate lab located just down the hall from Wold's university lab in space Wold rented from the university. The financial relations coupling the two were intricate. In addition to supporting Wold's basic research, the federal government also supported his corporate research, as did a biotechnology firm in Texas attracted by its therapeutic potential. Created in 1999—four years before SLU set up its conflict-of-interest committee—Wold's amalgamation of academe, business, and government violated no university regulations then in existence and in its first several years drew notice only for its scientific work.

But meanwhile, as described in part 1 of this book, the rules for academic-commercial relations were rapidly changing nationally in response to the Gelsinger episode and other misdeeds and abuses that evoked public and professional scorn and institutional embarrassments. SLU, a scandal-free onlooker, realized that the old easygoing ways of linking science to business were going out of fashion. Under pressure from the federal agencies that paid for science and stricter codes of conduct proposed by professional associations, the fire walls were going up and the bright red lines were being laid down at research universities throughout the country. Though they allowed for dealings between academic scientists and commercial enterprise, the new rules called for restrained intimacy, more—if not full—transparency, and closer monitoring by university authorities. Relationships that once easily passed muster were now suspiciously eyed as possible conflicts of interest. At SLU, the laissez-faire attitude that prevailed when Wold established his commercial laboratory on campus was overshadowed by concern for the university's public image. In 2001 Wold, with his easygoing candor, discussed his academic-corporate arrangements at a public conference. To the dismay of image-conscious administrators at SLU, Wold's remarks showed up in 2005 in a book critical of academic-industrial dealings, *University, Inc.: The Corporate Corruption of American Higher Education:*

> Wold introduced himself as the president and CEO of a small start-up, Virex [*sic*], which is trying to develop new genetic treatments for cancer. The company, he noted, was being run out of his own academic lab on campus. "When I started this whole business, I was an academic scientist and I was really motivated by the idea of bringing biotech money into the department," he explained. "I thought that I could handle both being a scientist and the president of a small company, that there wouldn't be a conflict of interest. But, of course, that was wishful thinking. It turns out that my activities with Virex *do* interfere with my duties as chair [of the department] and as a scientist mentoring students." He continued: "I'm sure that every one of you is thinking, 'Is this really a proper thing to do?' I don't know. . . . I rationalize this by thinking that in the long run this is going to be good for everybody because our goal is to develop an anticancer drug. So I try not to worry about the conflicts of interest."

The author of the book then related that "during the question period, I asked Wold how he planned to handle the intellectual property in cases where graduate students and postdocs were assigned to work on company-related research. 'We really don't have rules at our university to deal with that,' he admitted."[2]

In his conversation with me, Wold confirmed the accuracy of the book's account, explaining that time spent working with his firm would necessarily reduce time available for his university duties. He also pointed out that at the time, SLU had not adopted regulations governing intellectual property rights for students and postdocs working on company business. With his ingenuous manner, he appeared puzzled that these matters had aroused concern. Indeed, they had. The passage suggested profit-seeking at the expense of traditional academic values and indifference to the rules of the game. Embarrassment now intruded on SLU's long-standing pride in the scientific accomplishments of Professor Wold. The organizational and financial arrangements that once went unnoted became a matter of concern.

For the uninitiated in academic-commercial relationships, Wold's linkage of the two realms may appear incomprehensible. Funds from several government programs supported research in both his university and corporate laboratories. At the same time, contractual ties between his own company and still another company provided him with additional funds. He recycled part of the money received from this other company back to the university for research in his university and corporate laboratories, while he personally took part of those funds as salary for his corporate work. Labyrinthine as it may be, deals of such complexity were not unusual in the go-go days of academic-industrial cohabitation—and it's likely that many still exist without drawing attention. SLU officials acknowledge that they were slow to adapt to the new environment for academic-commercial dealings. Wold's misfortune was to be caught in midstream when the rules changed.

I met with Wold in an office adjacent to his university laboratory. Open-mannered and easily responsive to questions, he conveyed the impression of a man beset by difficult circumstances and pleased to find a visitor interested in his plight. He explained to me that his research employed genetically modified common cold viruses, so-called adenoviruses because they are isolated from adenoids, to invade, infect, and destroy tumor cells without damage to surrounding healthy tissues. The research was progressing satisfactorily, he told me. "We have a lot of good data in cell-culture systems constructing these viruses and characterizing them in cell culture and a lot of good data in animal models, where we can do a good job of suppressing tumor growth in animals. We've had some discussions with the FDA, and they've indicated what sort of experiments they want us to do with our sort of main product." In addition to receiving basic research support from the NIH, he said, he also hoped to receive a grant from a special program of the National Cancer Institute for drug development by academic researchers. For conducting basic research, he said, he was a university professor, but for drug development he was a corporate CEO.

"Don't the roles overlap?" I asked.

"They do and they don't," he said. "The company is the natural outgrowth of my NIH-funded research. I've been funded for over twenty to twenty-five years or so from the NIH, primarily what they call RO1 grants [for basic research by academic investigators]. On adenoviruses, I'm an adenovirus molecular biologist." For research by the company, he explained, "I rent a lab down the hall from the university. The corporate lab is clearly distinct from the university lab. I have a formal agreement with the university where I pay rent. It's a 250-square foot lab. I pay $13,000 a year, which is above market rate actually." His company, VirRx, he said, employs four PhD's and an animal research technician.

"You're wearing two hats obviously. You step down the hall, you're CEO. When you come back in here, you're a professor and chairman," I said.

"Right. That's correct," Wold agreed.

"These are very closely associated roles, aren't they?"

"The goals of the company are somewhat similar to my research program as a university professor. That's true," he agreed.

"Sounds to me like an excellent arrangement," I said. "It certainly saves a lot of time, doesn't it?"

"It saves a lot of time," he said. "It's a very convenient arrangement for both the company and the university. It's a very good deal for the university," he stressed.

"Why is it a good deal for the university?"

"Because, up to this point, all the intellectual property that has been developed belongs to the university," he replied. "And all projects that have been conducted by the company have been collaborations with this."

"So the university gets the patents even for work that's been developed in the company laboratory?" I asked.

"I take the position that my brain mostly belongs to the university," Wold said. "I don't have any unique ideas at home in my garage on weekends. And so as a result, the university has a lot of intellectual property that has come out of the project. We have three issued patents, probably four issued patents—one is just about to be issued. And we have quite a large number of pending patents."

"All owned by the university?"

"By the university," he answered. "So what my company does is license this intellectual property from the university. I pay a license fee. I paid an initiation fee for that license, and I pay an annual maintenance fee. And then if we're ever able to develop a drug, the university will receive royalties."

"Which could probably be quite substantial," I suggested.

"Well, if we can go all the way to an actually commercially licensed cancer drug, it should be substantial." But then with modesty not often evident on the frontiers of biotechnology, Wold added, "The odds of doing that, as you know, are actually pretty rare. But we're doing pretty well so far."

As our discussion proceeded, Wold's description of his academic-commercial relations entered deeper realms of complexity.

"The way I have this set up," he explained, "my company actually gives the university a grant, a formal grant—it would be similar to an NIH grant—and we pay indirect costs to the university."

"You give the university a grant?"

"The company gives the university a grant," Wold said.

"Where does the company get the money from? There are no revenues, are there?" I asked, recalling that the company had not yet marketed a product.

"The company has two sources of money. One is I formed a partnership with another biotech company, the company's name is Introgen Therapeutics, Inc. They're a Texas company, a cancer gene-therapy company. And so they like the technology we've developed, and so they invest in our company. So I receive money from them every year."

"How much is that?"

"I'd rather not say. Let's say it's a substantial amount," he said.

"And then you give a grant to the university."

"I give a grant to the university," Wold repeated. "And in return for that money they're putting in my company, they gain exclusive rights to this intellectual property."

"The university does?"

"No," he instructed me. "Introgen does. So, I license the intellectual property from the university, and then Introgen licenses it from me. This also is a very good deal for the university, because that provides a large infusion of cash into the research programs of the university." The other source of money, he explained, is from two government-wide granting programs, via the NIH in this case, designed to promote technological innovation by small firms: Small Business Technology Transfer, abbreviated as STTR, and Small Business Innovation Research, known as SBIR. "We have about six or so of those grants," he said. "Most of those are Phase I grants, which are of the order of $100,000 to $200,000 per year for one year typically. We've also had a Phase II grant, which was about $700,000. And then we're pretty close to another Phase II grant. So our research income a year is of the order of about a million dollars a year from all sources for the past two or three years. As I said, it's a good deal for the university because that money is all R&D money, and the research program is really the university's program."

"What about your people in your laboratory here or down the hall in your company laboratory—postdocs, graduate students? Where do they fit in?" I asked.

"They are funded by the university, by the grant that VirRx gives to the university. We call that a sponsored research agreement. They're funded by that."

"Postdocs and graduate students?"

"I have one graduate student who works on a project."

"How many postdocs do you have?"

"They're not postdocs," Wold said. "They're university faculty—research faculty. They could be postdocs. I have a postdoc who works for VirRx."

"So you have research-track faculty. They belong to the university, but they're funded by—?"

"They're funded by the sponsored research agreement that VirRx gives to the university. And they're also funded through these STTR grants. All of these STTR grants that we've submitted, and also SBIR, there are two budgets. There's the VirRx budget, and there's the Saint Louis U budget. The applicant is VirRx, and the subcontractor on the grant is Saint Louis U. The Saint Louis U part of the budget is exactly like an NIH RO1. There are full indirect costs that are paid to the university."

"The applicant for the SBIR grant is VirRx, and SLU is—?"

"SLU is a subcontractor," he explained.

"Is this a smooth-working arrangement?" I asked, trying to conceal my befuddlement.

"I think so," Wold assured me. "And fair. And so, for how I see it, everybody wins. My company wins, Introgen wins, and SLU wins. And you asked the question, is there a sort of intermingling of this? I would say the intermingling is that this is Bill Wold's research program, and it's my research program as a SLU faculty member. And it's my research program as a VirRx CEO."

"Substantively, the two research programs are quite similar?" I asked.

"They're quite similar. And what it does is it accomplishes the aims of the STTR-SBIR program, which is the transfer of technology from an academic lab into small business. It satisfies the aims of the biotech company, VirRx, and of my biotech partner, Introgen, in that it's a drug-development program, and it satisfies the aims of Saint Louis University, in that it's a fundamental research, translational research program."

An Unusual Arrangement?

"But I think we have to acknowledge that it's a bit unusual," I timidly suggested.

"I don't know how unusual it is," Wold countered.

"Are there others you've come across of a similar nature?"

"I wonder whether all university spin-off companies aren't similar to this," he said, hesitating for a moment, and then adding, "I think they probably are. I haven't had really close conversations with colleagues at other universities, but I think all university spin-off companies are probably very similar. I mean, they are university faculty who have research programs, and they spin out an idea of their regular research programs. What may or may not be unusual is just the precise way we've set this up."

"You mean just having it on the premises?"

"Just having it on the premises might be a— But we don't intend that that be a permanent situation. It's been here since 1999. But we're still at a very embryonic stage. It's a very small company, and it's all R&D. I think if we are to get into the next stage, which is through the clinical trials, then that will be different. Then we will move off; there will be a lot more money involved and actually different kinds of research involved."

"Do you derive a salary or fees for consulting or for anything?" I asked.

"As president of the company and CEO of the company, I pay myself a salary, which is on top of my university salary. Also with these SBIR and STTR grants, there is a 7 percent fee which is associated with those grants."

"Fee paid to whom?" I asked.

"Fee paid to the company. You can think of it as a profit, but I don't touch that money. That money sits in the company's coffers and will be used for R&D or will be used for whatever I decide to use it for."

"How much salary do you pay yourself?"

"It's about $51,000," Wold said without hesitation.

"You're aware of the fact that there was a book that came out awhile ago; it's called *Universities, Inc.*"

"I'm quoted. It was just a very modest meeting," he said, referring to the gathering where he made the remarks about his research arrangements at SLU. "I think it was the Midwest Graduate Association, and I was invited to speak along with a fellow from Wash U, and it was just about how I set myself up and basically what I'm telling you, except it was at an earlier stage."

"But the way it was presented in the book, there was a critical slant to it, as though you were violating some rule."

"That's the way it seemed to be that the author of that book presented it," Wold said. "That certainly wasn't the way I looked at it. I don't think what I'm doing is unusual or unethical or immoral. I think it's very straightforward. And I was very honest about how it works. I think I was asked a question: Is this compromising my ability to function as the chairman? And I said it was, but what I meant was any research program—I used to have four RO1s [grants]—and running four RO1s and performing my administrative duties as a department chair are in conflict. You know, just everything you do is in conflict."

"The commercial aspect was not really in conflict with the academic?" I asked.

"It's only a trivial amount of time anyway," Wold insisted. "I mean, the amount of time I spend on business, as it were, is virtually nothing. I have a director of operations who works for the company, and all the kinds of business things, details, are handled by that person. So I spend all my time thinking about research."

"And it's your sense that just in terms of what goes on with universities and business and science, this is not anything unique?"

"It's certainly something that has the potential for being a conflict of interest," he conceded. "And I think these kinds of arrangements in universities have to be examined very carefully. I would submit, though, that my particular arrangement is a win for everybody. As a department chair, I have the potential to be exploiting the space in my department. But I don't think I do. Nobody complains, for one thing. I pay above-market-rate rent. And the university reaps all the benefits. All the intellectual property belongs to the university. We publish our papers."

"You publish everything?"

"We publish everything," he said.

"You don't withhold anything for proprietary reasons?" I asked.

"I mean, I might. You can imagine a situation where I might, because you don't want to publish until you have your intellectual property protected. But that's the same as you have for RO1 projects. The university wants you to write patent applications, even for RO1-funded research. So there's no difference at all. We go to meetings; I get invited to meetings. There's an American Society for Gene Therapy meeting in June here in St. Louis, and we submitted four abstracts to that meeting. The first author on three of those abstracts was my graduate student. It was a good project for her."

"This arrangement of having the company nearby, was this approved at some higher level? Is there a committee or a dean or a provost or somebody?" I asked.

"There wasn't a committee when we set this up. Sure, it was all very transparent and aboveboard, and the dean of medicine—my boss—the associate provost for research, and the university attorneys and the provost, I presume, were all aware of this arrangement."

"When you say they were aware, did they say, 'Okay, go ahead'?"
"Yes."

"So there was the general counsel and the deans and the provost and the president," I said.

"I don't know if the president knows about this, but he certainly could," Wold said.

"So, it was transparent, aboveboard. The general counsel, the dean, everybody knew about it."

"Yes. Just within the last year, the last year and a half or so, we have a Conflict of Interest Committee. They have become involved. And they have sent me questions," Wold said. "They wonder whether there's a conflict of interest in terms of my fiduciary responsibility, because I'm CEO of this company. They wonder whether the laboratory—whether it's contiguous with my academic labs. I don't know how you define contiguous," he said with obvious annoyance and upward palms. "It's a separate room."

"Why would it matter, even in the same room?" I asked.

"I don't know. In the same room you can see if there are some SBIR-STTR situations where they put a piece of yellow tape down the lab," he said, grimacing at the absurdity of the idea.

"What else was worrying them?"

"The VirRx part of the lease agreement that the VirRx employees can use the university equipment."

"Do they pay for that?" I inquired.

"We pay for that in terms of the rent. That's the deal as part of our lease agreement," Wold explained. "Those are the main issues. And then the other issue would be whether I'm exploiting the faculty and graduate students to work on this project. So, my argument is why don't you just come in and investigate whether they're being exploited or not. They're publishing papers; their careers are advancing; they're going to meetings. They're inventors on the patent applications, if it's appropriate."

"The Conflict of Interest Committee is just asking questions; they're just informing themselves. Nothing accusatory," I said.

"I don't think so," Wold replied. "I think we have to come to an agreement. Right now, they're questioning and I'm responding. I'm not sure how it's going to resolve. We may have to—there may have to be some changes in the arrangement. I'm not sure."

"What kind of changes might potentially be in the works?"

"I may have to not be a senior officer in the company."

"Would that be acceptable for you?"

"I'm sure I could make an arrangement," Wold said, "but I don't think it's a good idea, because I don't see that there is a conflict. I think what a senior officer in the company does at this stage of the company is direct the research program. So it would be sort of ridiculous. It's my research program. It would be ridiculous to go out and hire a biotech CEO to direct my research program. Maybe a few years from now when we're raising money from venture capitalists or something like that, then you would have to have an appropriate businessman."

"Or if you go into clinical trials, then the whole nature of this changes," I suggested.

"That's correct," Wold agreed, "because that's not my area of expertise. Clinical trials would be done through a collaborator, a scientist who was an expert medically. I don't do clinical trials, and even if I did, there would be full disclosure, as there always is. Whenever I talk at a meeting or publish a paper, it's always clear that some of this work was funded by this company."

Wold now asked me a question: "You have a lot of experience in this area. What do you think of this arrangement?"

I replied, "You suggested that this arrangement is not all that unusual. I have not encountered one before like this. My impression is that SLU is behind a bunch of other universities in developing all these regulations and fire walls and safeguards, but that there's an awakening interest here. I think it's manifested in this Conflict of Interest Committee, and therefore your arrangement attracts attention. People are anxiety-ridden about the potential for bad publicity," I pointed out.

Wold replied, "I hope they examine the specifics of the situation, rather than from the blanket rules where they are more worried about potential than reality. Even if we were in clinical trials, I still don't see anything immoral about this. What is immoral about me wanting to take my drug and do a clinical trial? That's exactly what I want to do."

"The stock answer regarding clinical trials," I responded, "is that people who have an interest in the outcome of clinical trials should be hands-off."

"But everybody does it. That's how it works," Wold said. "So as long as you're honest with your data and you publish your data, including your negative data."

"You're trustworthy," I replied, "but some people figure the next guy may be so eager to get to market that he overlooks some negative data."

"Let's say I'm Pfizer," Wold said. "That's what they do. It's their data. And it's their data that's used to support the clinical trial. So I don't see how that's any different from what I'm aspiring to do."

I turned to another topic. "Why was it advantageous to form the company instead of keeping it all in your SLU laboratory?" I asked.

"Because it's a drug-development issue," Wold replied. "And that means it's $100 million, $500 million, whatever that number is, to develop a drug. So at some point it's going to have to go into the commercial arena. And so there are two ways to get at that. Either through regular NIH RO1 grants, and then you license to a company. Or I develop my own company. In either case, it has to go into the commercial arena in order to have access to drug-development money."

"But the conventional way," I observed, "is that it goes from NIH-financed research in an academic laboratory to a license to a company, with the university holding the patent."

"That's true," Wold agreed. "But it's almost part of the career pattern of a successful university faculty member to develop spin-off companies. I know many people who have developed spin-off companies, and that's what they do. And I had experience actually with the other way, where I had an arrangement with another company who had been licensed by the university to develop technology and that was a big disappointment, that arrangement. The reason that is, when another company licenses your technology, what you have is an advocate in that company for your technology. And you're completely dependent upon the success of your advocate. If that advocate decides, he or she, if they get fired or if they move to another company, or if their bosses decide they don't like that technology, then you're out to lunch. So the way I did it, as long as I can raise money, then I'm the advocate. And of course I'm pushing it. That was a major aspect of it. And of course I'm not going to lie to you. The way I did it, I stand to make more personally. And I don't see anything wrong with that, either. This is America. It's more fun this way, too."

"I Would Love to Cure One Patient"

"This will get it to the bedside faster?" I asked.

"I think so, because I'll always be there pushing it. My technology is based upon the virus making a protein that is named ADP [adenovirus death protein], and I discovered that protein with my own hands back in the mid-'80s. Then working with other colleagues in the lab—an outstanding scientist named Ann Tollefson—we discovered the function of this ADP protein, and we came to the conclusion if we made a vector that made this ADP protein that this would be a good anti-cancer drug. So this really is a wonderful example of bench to bedside. This is a basic scientist playing around in his own little lab by himself in the evenings, making a discovery and then trying to take that very basic discovery in very obscure adenovirus molecular biology and trying to discover a drug from that discovery. So this is a perfect example of what you call translational research: converting a basic research finding into a drug. So if I could accomplish that in my career," Wold said, "that would be very actualizing for me, very satisfying, I could say I really did something.

"Another reason for getting into this company business," he continued, "is that many basic scientists like being basic scientists, that's why you go into it. But it's also frustrating, because you really want to help people, and you always say, my basic research is somehow going to get translated into a cure for cancer or something like that. But you know it probably won't, at least not directly. But here's a situation where I'm trying to make it directly. I would love to cure one patient. So that's a noble motive. I wish science were such that more basic scientists had the opportunity to develop drugs. It's very, very difficult. There's so much money involved, and it's so difficult, and there are so many roadblocks and rules and regulations. It's almost impossible. Most basic scientists don't even try."

: : :

In November 2005 SLU's Conflict of Interest Committee decided on a two-step resolution of the Wold dilemma. First, he would move the VirRx laboratory from its site next door to his university laboratory to another building on campus. And by June 30, 2007, he would move the VirRx laboratory to an off-campus location. Professor Kathleen Farrell, chair of the committee, told me that "the world has changed." There was "no suggestion of wrongdoing" in the committee's attention to Wold's business and academic relationships, she emphasized. But, she explained, "there needs to be a clear distinction between university business and private business." I asked Farrell whether Wold would be permitted to remain head of his company and retain his SLU professorship. She replied, by e-mail: "The specifics of conflict-of-interest management plans are confidential in accordance to the university policy. I can say that we agreed on a management plan which includes the exit that I discussed with you earlier. Dr. Wold has agreed to the management plan."[3]

Wold told me the new arrangement was workable. Then he said, "One year I'm the paradigm of an entrepreneurial faculty member. Next I'm in conflict of interest."

The new arrangement will fit the new rules. Whether the new rules are more conducive than the old to the development of useful drugs is not clear to me.

11 Profits and Principles

The University of California, San Francisco, was a trail-blazer in the biomedical revolution of the late twentieth century and the ensuing fusion of academic science and commercial enterprise. UCSF's Herbert W. Boyer, in collaboration with Stanley N. Cohen, of Stanford University, developed the technique of using enzymes to cut up strands of DNA and splice them together. Their innovation, first published in 1973, created the technological basis for the thousands of companies, large and small, that constitute the worldwide biotechnology industry. UCSF, along with Stanford, became a spawning ground for scientific-industrial collaboration. The deals took many forms, including start-up firms founded by university faculty; advisory posts for professors in biotech and pharmaceutical firms; and industrial consultancies for academic scientists, courted by companies, both for their professional skills and reputations that impressed Wall Street. The biotech firm Genentech, cofounded with $500 each by Boyer and Robert Swanson, a venture capitalist, was one of the first start-ups to exploit the new science. The firm quickly became a giant of the fledgling industry—though in circumstances that presaged that the combination of biotech and business would not invariably follow the collegial traditions of academic science. The role of brazen corner cutting in Genentech's early success became clear in U.S. District Court proceedings in 1999, when the University of California sued the firm for violating its patent for producing human growth hormone. A former Genentech employee testified that in 1978 he secretly removed from a laboratory at UCSF a bacterial clone that helped lead to Genentech's patent for producing the lucrative hormone.[1] After the 1999 trial got under way, Genentech settled the case with a payment of $150 million to UC and a $50 million contribution to construction of the first research building at UCSF's Mission Bay campus. The sums were bearable for Genentech, given that at the time of the settlement, Genentech's total sales of human growth hormone products exceeded $1.2 billion.

The Biotech Boom

The commercial potential for biotechnology and its leading firm, Genentech, was apparent from the start. In 1980, within minutes of the first public offering of Genentech stock, the price rose from $35 to $89 a share. Twenty-five years later, following numerous stock splits, a hundred shares purchased at the initial offering price were worth over $600,000, and Genentech, with annual sales of $3.7 billion, ranked second in the biotech industry, behind Amgen. The biotech boom was on, and at the center of it was UCSF, unique in the California system of public universities because of its exclusive focus on the life sciences and health professions: medicine, dentistry, nursing, pharmacy, and a graduate division of training and research in basic and clinical sciences and other health-related fields. From way back a powerhouse in biomedical research, UCSF prospered under the NIH competitive award system for research money, which ensured that the strong get even stronger. In 2005, with a student population of only 2,800, UCSF's full- and part-time faculty and staff numbered 18,600. The annual budget, $2.3 billion, included over $700 million for research. UCSF ranked fourth nationally in volume of dollars received from the NIH, and nearly seventy companies in the life sciences were directly linked to UCSF by inventions or faculty connections.[2] With scarcely any regulatory guidelines strictly observed in the 1970s and 1980s, entrepreneurial activities proliferated at UCSF, producing fortunes for many faculty members with one foot in academe, the other in science-based commerce. It was only years later that UCSF, like other universities, awakened to business-based scandals in academic science and undertook close attention to conflicts of interest, conflicts of commitment, restraints on publication, and other now-familiar

concerns about proper behavior in the management and conduct of research connecting academe and industry. On the UCSF campus, where
thickets of commercial relationships had flourished with little ethical
guidance or restraint, the reining-in process proceeded slowly and con-
tentiously. Often heard was the complaint that cumbersome regulations needlessly thwarted profitable deals and retarded innovation. The
difficulties of ensuring right conduct at UCSF and its sister universities
in California were doubly compounded by California law and custom.
California's public universities were not conceived as sheltered enclaves
of socially disengaged scholarship. Rather, utilitarian motives under-
pin their costly support by California's taxpayers. The universities are
expected to contribute to the economic development and prosperity
of the state. At the same time, as California state employees, faculty
members are subject to strict regulations regarding outside business
deals, income, and that venerable bête noire of public service, conflicts
of interest. Over a quarter of a century after the Cohen-Boyer discovery, the development and application of new rules were still under way.
Lisa Bero, professor of clinical pharmacy at UCSF and chair of the
Chancellor's Advisory Committee on Conflict of Interest, discussed
these matters when we met on campus on January 25, 2005. Bero initially conducted me on a tour of the complexities involved in parsing
closely related academic and commercial research when a faculty member is scientifically and financially involved in both.

"The biggest conflict I see with tech transfer," she said, "is investigators starting their own company. Because basically, what happens at
the University of California is if you're a professor here and you invent
something, the university owns your patent. But then what happens is
the investigators can't find a company interested in developing their
invention, and so often they go out and they found their own company,
get some venture capital. And so then they want to try to do that research, and the problem often comes with the company funding the
research they're going to do."

I asked whether they can research company problems in their university laboratory.

"No, they can't," Bero replied. Her chairmanship of the conflict-
of-interest committee, she explained, "gives me a skewed view of what
goes on here at UCSF. You're probably familiar with these committees;
every university has them. But basically, all they do is review the disclosed financial ties of faculty if they're getting funding from a company
that they also have a personal financial tie with. So, if somebody gets
$15 million from their own company, but they're not actually getting

any personal income in terms of consulting fees or anything from the company, our committee never sees it. And it's actually very hard to get an estimate on how much that might be happening. I think it's actually fairly rare, because usually you're getting some personal income from a company if you're a founder or you're on the board of directors or you're one of their major consultants. And then if you do, then our committee sees it. So that's basically what we deal with, investigators who are getting funding from their company and also have a personal financial tie in that company. The question for the committee is, one, is there a conflict of interest? And, two, if there is, how should it be managed? They can't do the *company's* research, but they can get money from the company to do their *own* research. And so that, of course, is the question. And that's now something that these committees grapple with. Do we think the investigator is actually doing the company's research in their lab, which they shouldn't be doing, or do we think they're actually getting money from the company to develop the next new invention?"

Separation of Academic and Company Research

"We've been actually doing a study of conflict-of-interest committees in the UC system," Bero said, "because what's interesting about that is they all have the same policies, but they all implement them in different ways. One of the findings from the studies is that one of the things these committees look for is separation of activity—that's what we call it. And so basically what they're looking for is if the activity the investigator is doing at the university is truly separate from the company. And sometimes it's actually really easy to figure out. Say somebody has developed in their lab some peptide, and it's a peptide to treat, or potentially to treat, Alzheimer's. So they start this company with venture capital that's going to develop this product. And the investigator is actually really interested in studying memory. That's kind of what they do at their basic science work. And so the company is working on developing that patent they got five years ago into a product. Meanwhile, the company also gives the investigator a research grant to look at memory function in mice. Nothing to do with that patent. It actually may be to develop the next new thing. So the potential conflict there is the company's funding them to do some basic science work that could eventually maybe come back to their company. But, you know, these committees say, 'Well, that's separation, because actually what's they're doing in the lab is very different than what the company is doing.' "

"It sounds like there's a lot of overlap," I said.

"If one's applied and one's not applied, strictly lab research, that's separation. Sometimes there is overlap, and in those cases, the committees usually don't allow it," she explained.

A Reluctance for Oversight Duty

"Are there many occasions when they don't allow it?"

"On rare occasions, the committee actually recommends not to accept the funding whatsoever. In other cases, they recommend some sort of management strategies. They'll say, 'There's a conflict of interest, but we'll appoint an oversight committee that will monitor this investigator's research.' Our campus actually isn't too keen on oversight committees. Other campuses are. One response from faculty members: 'Why should I spend a lot of my time overseeing another faculty member so they can make a lot of money?' It's actually really hard to convene these committees. And it's a lot of work, because you really have to get into the science, and I've been on one of these committees, and it's like you have to interview all the postdocs and the graduate students, and so the other thing is actually to remove the investigator from a decision-making capacity related to the grant. So, basically, they're not the principal investigator anymore. That kind of thing."

"Postdocs and grad students: What are they allowed to be involved with?" I asked. "They can't be doing the work of the start-up in a university laboratory, can they?"

"No, they should not be doing that," Bero answered. "And actually one of the themes that's come up in our analysis with the committees is—one of the questions people are usually asked in their disclosure is, 'Are postdoctoral students involved in the research?' And if you check yes, that's kind of a red flag to the committees, because the committees are very concerned about students getting involved in company research, not for the conflict-of-interest reason particularly, but because they're really concerned about students not being able to publish and get their dissertation in a timely manner if there's some holdup with the company, with data. So, basically, if students are involved in the research, they have to get uninvolved, and faculty have been told you can't have any students and postdocs involved in this basically."

"Is there also concern that the student will be possibly working on a rather narrow, applied problem rather than being involved in basic research and getting a broader view of the scientific field?"

"That doesn't come up on these committees," Bero replied, "but actually it's sort of talked about obliquely quite a bit. But we don't actually get into critiquing whether somebody's research is useful or not or are they a good mentor."

I noted that several faculty members had told me that, in contrast to the early days of the biotechnology boom, the present-day UCSF culture puts greater emphasis on public service and was tilted against commercial involvement. "Do you share that impression?" I asked Bero.

"UCSF," she said, "is different from private institutions in that we have these state disclosure guidelines that are very low. Basically, faculty have to disclose personal financial ties that are above $500. Until a couple of years ago, it was $250. So basically they have to disclose every time they write a grant related to a company virtually all their financial ties. So there's a much higher level of scrutiny. More public-spirited? I don't know. There're a lot of people here who are keen to work with industry. It is a public institution, and a lot of people would go to work somewhere else if they weren't interested in working in a public institution. We also have a very strict clinical trial policy, which is different, that is very different from other campuses. Basically, we have a policy that says if an investigator is getting funded by a company to do human subjects research—that's broadly defined—then during the course of that trial or study they're not allowed to have any personal financial ties with that company. It's a zero limit. So, they can't have honoraria, nothing."

"But the company may fund their research?"

"Oh, yeah. And there's a lot of that here. But other universities, if they have any limit at all, it's usually set at $10,000. So they basically say, it's okay for the investigator to get $10,000 in consulting fees from a company if they're also getting research funding from that company, and we would say, no, they can't get anything in consulting fees."

"What if they donate the consulting fee?" I asked.

"Doesn't matter," Bero said. "And we have something here called a compensation plan. So, actually, even when we get consulting fees, most of them go to our department anyway. But we still count those as a personal financial tie, even though 70 percent of it is going to your department. It's still not allowed."

Clueless about Conflict Policies

"Is the conflict-of-interest situation more or less under control?" I asked. "You seem to have established techniques for dealing with it. Committees in place. There must be a fairly highly level of awareness of what the rules are."

"We've done a system-wide survey of faculty, and we've also done some in-depth interviews of faculty both at here and Stanford, clinical faculty only. And one of the things we asked about is people's aware-ness of the conflict-of-interest policies. And basically, unless they've run up against the committee, they've no idea; they're clueless. The way you find out about these rules—generally it's not part of the faculty orientation. It's not something that chairs talk to their faculty about when they're hiring faculty. Generally, the way you find out about it is you're writing a grant; you're under this huge deadline and pressure to get your grant in. Somebody throws this form at you and says, 'This is the financial conflict-of-interest form—fill it out.' You fill it out, and the next thing you know, all of a sudden you're getting reviewed by this committee. And you're like, where did this come from? So, it's really that a lot of faculty don't know about it. I do think that's changing a little bit, though, because I've actually been contacted as chair of this committee by chairs of other departments who are recruiting faculty to anticipate financial-ties issues, and they want to meet with me to learn about the policies. Like what are they getting into if they come here. So I do think the chairs are becoming more aware."

Rules Too Strict?

"Do you ever come across a case where someone says, 'These conflict-of-interest rules are a little bit too strict; I really have some outside business interests I want to pursue in addition to my academic duties.' And not come here if you're trying to recruit them?"

"I've never run into anybody personally," Bero said. "There's sort of an urban myth around here that our clinical-trial rules are so strict that we've driven faculty away by the droves. But I actually looked into it, and that's not true. I only know one faculty member who's left, purportedly for that reason. Also from when we did our interviews with faculty, my favorite quote is, 'Conflict-of-interest rules are just a minor cog in a wheel of abuse at this university.' I could say that as an investigator, you have to put up with so much paperwork, like what's another rule? Big deal. It's not going to drive me away, because I want to work here. The other thing is we have a lot of investigators here who don't have these kind of financial ties that will come to the committee. Again, I'm talking about personal financial ties."

"They're not consultants to outside companies or start-ups?" I asked.

"They may be a consultant, but they're not a consultant to a company that's funding them. Or they may get funding, but it's not one they're consulting for. There's no evidence at all that it's turning people away."

I noted reports of prospective faculty members inquiring about rules for royalties on licensed patents and other commercial matters.

"Yeah," Bero acknowledged. "I want to think that's why more faculty are being sent to me. I would imagine that would be true. The other thing is we have had some more senior faculty who come from an environment where, basically, if you have an invention, you keep the patent for yourself. Now, all of a sudden they're here, and they go, 'What! The U of C keeps the patent. I can't believe it!' And so that's been a little shock. Here, basically, whether there's federal money or not, you can't keep the patent."

Many without Company Ties

Recalling UCSF's reputation as an entrepreneurial university, I asked about the extent of faculty involvement in commercial activities.

Bero replied that "about 30 percent of the faculty—it fluctuates by year—have any sort of financial tie that meets our disclosure threshold. That leaves 70 percent who don't. So that means 70 percent don't have these personal financial ties that conflict with their federal or their private funding. The FDA and others, they're always saying they can't find unconflicted people to serve on committees. My argument is, look at UCSF; it's not like we're all slackers. A lot of us don't have financial ties."

"Is it another urban myth that half of all professors in the life sciences at major universities are in one way or another tied up with commercial enterprise?"

"One way or another, that might be true," she said, "because I'm talking about these things that we define as conflicts. And we do have a lot of faculty here who get like a ton of money, say, to do clinical trials from companies, but they just don't do a lot of outside consulting. So, one way or another, I think there probably are a lot of ties with companies."

"When they get a ton of money to do the clinical trials for outside companies, does any of that adhere to them or is it simply for the cost of doing the trials?" I asked.

"It's simply for the cost of doing the trials."

"They don't personally profit from this?"

"Correct," she said.

"Is this a situation that is essentially under control," I asked again, "or do you find eruptions of unseemly behavior here and there with some frequency?"

"I actually think it's pretty under control. We haven't had any scandals lately. All the big scandals we have had had to do with people who are getting their rights to publish suppressed for one reason or another." But that issue goes back sometime, Bero noted, referring to several well-publicized cases at UCSF in the past decade or earlier.

Exemptions from the Rules

I asked Bero whether she foresaw new problems from recently enacted, well-financed California state programs to promote academic-industrial collaboration, particularly at UCSF's new Mission Bay campus.

"Yeah," she replied, "a lot of new problems arise. I don't know even how they're managed, because they don't necessarily follow all of our regular university guidelines. They get all these exemptions. A good example is these STTR and SBIR grants, which are basically where an investigator has to collaborate with a company to get the grant. And a lot of times investigators wind up collaborating with a company in which they have an interest. And even the federal guidelines say our conflict-of-interest guidelines don't apply to these, because how could they? You're basically putting this faculty member in conflict. And so, there are exemptions made for those kind of grants. They're not viewed in the same way."

"The university grants an exemption?" I asked.

"No, the federal government," she said.

"And that overrides any university rules?"

"Yeah, they say you basically don't have to manage these like you would manage a regular NIH grant," Bero responded.

"But you're managing conflicts of interest here because of the university's own values, not simply because the federal government says you must do this or you may not do that. Doesn't this override raise local concerns?"

"It does raise local concerns. I can tell you that," Bero assured me. "On the conflict-of-interest committee, it certainly has. But the problem is evidently we don't have the authority to actually impose a management strategy if the federal government says it's okay, as long as they're aware of it. They know about the disclosed financial ties if there is one. As long as they say it's okay, then we can't say you can't take that grant or you can't do this or you can't do that. But I do know there've been concerns among faculty. And we actually have put restrictions on clinical trials. It's only happened like once or twice," Bero said, but declined to discuss the details.

Bero told me that researchers at UCSF who are funded by the Howard Hughes Medical Institute, the wealthiest of all U.S. philanthropies for medical research, are "not under UCSF rules. They're their own little operation. I don't quite know how that works," she acknowledged.

"But Howard Hughes investigators also hold university appointments, don't they?" I asked.

"Not all of them," Bero said. "And they can also submit grants through Howard Hughes and not UCSF. Like, for example, we have a lot of faculty who have an appointment at the VA and at UCSF. And the VA has this big clinical trial in a center, and some of our faculty say the VA rules are actually less stringent than UC, so when I write a clinical trial, I'm not going to submit it through UCSF, I'm going to submit it to the VA. So, it's the investigator who can choose where they want to submit it through. I'm in a couple of departments, so I can submit my grants through Department A or Department B. They're both under the university, but if I was in another institute, then I could say, 'Well, I'm going to submit it through QB3 [a California state industrial innovation program] or the VA or UCSF.' If it's through the VA, our lawyers never need to see it."

"You've been skeptical about the value of disclosure," I observed. "Is that because it lulls people into complacency about a deal, making them think it must be all right because the financial connections are openly stated?"

"I think disclosure is really good about managing perceived conflicts of interest," Bero explained. "If universities' main concern is really perception, if anybody's main concern is really perception of conflict of interest, I think disclosure is a good thing. Get it out there in the open. But there's very good evidence that disclosure does not reduce bias. I've done a lot of studies looking at the association of funding and outcomes, and basically finding that studies that are funded by a particular type of company, compared to those that aren't funded by that company, tend to get outcomes that favor the company. When I started doing this, nobody had actually proved that. And now there's my reams of data. You do this research based on disclosed information. So basically you find published articles, and the published articles have like disclosures of who funded them or financial ties of investigators, and then you just classify the outcomes as whether they're favorable to the sponsor or not. Like if it's a drug study, it's very easy. Favorable means it finds the company's product is more efficacious or less harmful than competitor or placebo. That's considered favorable. So you just compare that and you can also control for whether it was peer-reviewed, what year it was published in—you control for lots of stuff. All of that data is based on disclosed funding sources. So obviously disclosure doesn't prevent the bias. And there're a lot of reasons for the bias, obviously. So that's why I think disclosure doesn't prevent bias. So for people who are concerned about conflict of interest because it might cause bias, disclosure is not going to do it. If they're concerned about perception, then fine, disclosure is a good thing. There's very little research actually on what your average member of the public thinks of financial ties of researchers."

"Do you get any signals from federal agencies about conflict of interest?" I asked.

"I've been at a lot of NIH meetings on conflict of interest, and some of those have been together with the Office of Human Research Protections. The signal they send me is concern; they're very concerned. Some of these meetings turn into, 'Well, whose responsibility is this, really, to manage these conflicts? Maybe NIH shouldn't be regulating this. It should be done at the university level.' Then we have our faculty saying about themselves, 'Well, it's really up to me; if I was in conflict, I'd know it, and I can handle it.' I think that's really the question: Who should be trying to manage this?"

"Is the principal concern in Washington that there are some mechanisms in place that are supposed to manage it, even though they may or may not be doing it well?"

"No, I think their principal concern has been that there have been instances where there's been bias in studies that seems to be related to financial ties, and how did that happen? And who should have prevented that?"

Bero explained that in addition to the conflict-of-interest committee at UCSF, there is also a Committee on Human Research, elsewhere known as the institutional review board. "They're separate committees, which are in communication," she said, "so if we're dealing in the conflict-of-interest committee with a study involving human subjects, we feed back our recommendation and everything to the Committee on Human Research. And at a lot of universities, the Committee on Human Research has been handling all of the conflicts of interest—they don't have a separate conflict-of-interest committee. I've seen some turf wars break out at meetings: the Committee on Human Research people think they should be handling the conflict-of-interest reviews. At UCSF it's not really practical, because a lot of our conflicts don't involve human subjects at all. Investigators can be doing pure basic science research and have a conflict of interest. So the human subjects committee would never see it, but they sure have a conflict of interest with what we were talking about earlier, with tech transfer, or they could be doing purely animal research and be an owner of a company."

"If you were in a position to implement improvements in the system, what would you recommend to make this work in a better fashion?" I asked.

"I'm actually finding rather disturbing that most faculty aren't aware of even our policies, unless they're personally affected by them. It sounds like a silly solution, but I actually think faculty need to start thinking about this and be educated about conflicts of interest and their potential for actually affecting the outcome of research. I think if faculty invested more in the process, there would be better management all along. In certain situations, we need to have more separation of funds. Certain things just shouldn't be allowed. I'm a radical in that regard. One thing we found in our research is that stock is very troublesome to people. And other people have also found that lay members of the public are much more concerned if somebody's got stock in a company that's funding them to do research than if they get consulting fees. I don't know why. But that's what the data show, at least in terms of public opinion. So maybe that's an area where you just shouldn't be allowed to—and I'm not talking about mutual funds. I'm talking about stock where you actually have control of the stock. Maybe that's an area where you just shouldn't be allowed to have interest over X percent in a company that's funding your research. You have to disclose, for federal grants, 5 percent equity or over, and for the state, I think it's much lower, $3,000 or something. But then the question is it's not banned over that level. You have to disclose it to the institution and then the institution decides what to do."

I asked about reactions at UCSF to revelations that some senior NIH administrators held lucrative consulting contracts with private firms.

"I think there was a lot of anger: 'Oh, we've been operating under these really strict rules, and look what they're been getting away with,'" Bero replied. "I find it fascinating that now they've banned these consulting agreements, that's sort of like our clinical-trials policy. We can't decide—what's an acceptable level of financial ties. Is it $500, is it $1,000? At NIH they just banned it, so that's sort of a precedent for universities that are starting to think about certain financial ties as unacceptable."

"Do you think that's desirable?" I asked.

"I think for clinical trials it is. I totally support the policy we have here, which is a total ban on personal financial ties with a company who's funding your trials. And I don't think there's any lower limit. How can you set a number for which there's no bias, if you do think there's a chance for bias?"

12   Generations Apart

The University of Wisconsin–Madison is the home of a renowned technology-transfer organization that we encountered earlier in this book, the Wisconsin Alumni Research Foundation (WARF). Founded in 1925, WARF has returned more than $800 million to the university from some sixteen hundred U.S. patents derived from research in campus laboratories. WARF currently provides $55 million a year for research at Madison, a welcome sum at any time but especially in a period of tight budgets for higher education. The humanities as well as the sciences benefit from WARF. Some of the money is for start-up grants for young faculty members who have not yet obtained outside research support. As a source of capital for over thirty companies spun off from the university, WARF contributes to the state's economy—a mandated role common to public universities. But the success of WARF and the gratefully received financial assistance it provides tend to obscure concerns about the possible negative effects of profit-seeking in an academic setting. Behind the beguiling statistics, the concerns are nonetheless present, as became evident in my conversation, on April 26, 2004, with Timothy Mulcahy, at the time a senior administrator at the intersection of science and commerce on the Madison campus. Mulcahy, a professor of pharmacology, was associate graduate school

dean for the biological sciences and associate vice chancellor for research policy, as well as chairman of the institutional review board. (In
February 2005 Mulcahy was appointed vice president for research at
the University of Minnesota.)

Wisconsin's dual commitment to traditional academic research values and commercialization of its research created tensions, Mulcahy
said, but, he added, "I think it's a healthy tension. I view it kind of like
pain, in a sense that pain is there to tell you there's something you need
to be attentive to. If you're not appropriately attentive to it, you could
seriously hurt yourself. If you are, you can navigate that problem and
hopefully come to resolution. So as soon as the pain is gone, as soon
as the tension is gone, that's when I get nervous," he said. "And what
I am somewhat concerned about—and maybe in a way this is difficult
for me, because in a way I'm baring my soul, and my colleagues might
be surprised to hear me say some of these things—what I'm concerned
about, and I've seen this happen, is that the young scientists who are in
training now, they're training in a different environment than I trained
in. I was trained in '75 to '79, just prior to Bayh-Dole, so my training
didn't include any of what I'd say is commonplace in university science
today, and that is the potential for patenting and licensing, the attraction to spin off companies. So that was not part of my training, and
what I see with new students is, it's very much part of their reality, but
I'm not sure that we've done an adequate job of informing them of the
potential difference in values or pitfalls and problems that might exist.
We haven't warned them enough about the tension. So they don't recognize the pain, as I've referred to it before, as readily, and aren't quite
as aware, to say, 'What do I need to do to resolve that problem?' So, I'm
concerned that they're growing up in a different environment."

"Are young scientists more attuned to commercialization? Are they
looking for it? Is that one of their goals?" I asked.

"I don't think it's one of their goals," Mulcahy replied, "but I think
it's a more commonplace and more recognized element of their world.
It's not unusual now to be working for faculty who have licensed and
patented technology or have spin-off companies. And that's not to say
that there's anything wrong with that activity. I'm just saying as it becomes more ingrained and a more common part of the experience, I
think you become less attentive to it. And to me that tension that exists is healthy, and as soon as you begin to ignore it or are not aware
of it, then I think you're setting down a road that could have some
consequences."

Declining Awareness of Ethical Issues

I asked why the acceptance of commercialization should arouse concern, noting that "the idea of moving knowledge out of the laboratory into the economy or to the bedside strikes many people as altogether desirable."

"I would argue very strongly that it is desirable," Mulcahy responded. "It's absolutely desirable. In fact, it would be a horrendous waste if, at the expense to the public, the research we did did not yield benefits for them in the long run. But, in that transition, I think we always have to be very attentive to what are the social costs, what are the ethical costs, and as long as we can, with a straight face, say we're doing the right thing, we're doing the best we can to balance that tension, the give-and-take that exists there, then I think we're okay. It's when we evolve the culture where that's no longer recognized as a potential contrast in culture that I think we run a problem. And I'm not sure about the new people coming in. Because they're coming into a culture that—I don't want to say is one-sided, because it makes it sound like it's a deliberate effort—but who may not be aware that some of these new issues might raise concerns about some of our core principles in universities."

"Are these abstract concerns that you're expressing? Or have there been incidents that give you cause for concern?" I asked.

"I think there have been some discussions that we've had, not at an administrative level, but at the core level, at the grassroots level that I've been involved with, which make me acutely aware of the fact that the more junior scientists, and that includes some junior faculty, are not really aware of some of the key issues and concerns. I attribute this to the fact that they have been trained and have worked in a system where this is very commonplace. Whereas for my environment, doing some of that commercial thing before was nothing that I ever thought about, and my major professor never discussed that. It was not foreign to me, clearly, because I was aware that it happened, but it was not something that I saw as a common part of doing science at a university. I think now it is viewed much more as a common element. Not everyone does it, but it's not peculiar or unusual to find people who are. And that's important, I want to emphasize. That's something I want to be sure to disclose to you, that I think that it is important that we do that. And I'm not one for closing the door, but I am one for being sure that people are aware of the issues that it raises."

Silence to Protect Patent Rights

"What issues does it raise?" I asked.

"I think the potential for conflict of interests. When you talk to people, everyone will admit, and I would agree, I'm not aware of anybody who deliberately subjugates their primary interests to their secondary interests. But I can't help but think that even unconsciously we are influenced by those factors, and even if we don't acknowledge it, you've got to wonder, does that influence what kind of experiments we do, what kind of patients you recruit to study. Now, the institutions have taken, or tried to take, a very responsible role in managing conflicts of interest, and it's a very steep learning curve for all of us, and I think we're doing a good job to address that. It's when we begin to ignore those kinds of issues that I have a major concern. That's one example. Being in a setting where new students, postdocs, are willing without challenge to agree to keep information in a scientific discourse confidential, such that we don't jeopardize the potential rights for patents or licensing."

"Is that actually happening in your laboratories?"

"I would say I was at one meeting where that was an issue. And to hear people more or less willing, in my opinion without really thinking about it, to say, 'Okay, I'll do that.' And have others, myself included, arguing, 'Wait a second, we also have to recognize science as open; science has to be open.'"

I noted that "pre the commercial era, a lot of scientists would normally keep cards close to the vest so they'd be first with publication."

"Right," Mulcahy said, "and I think there's still that. That's going to be innate; it's innate in our system. But what I'm referring to here is information that you would have in what I would call an open scientific meeting—at any one of our national organizations. You find people to be willing without questioning to sign a confidentiality agreement that anything you hear here you shouldn't talk about. That's a little disturbing to me personally. And to have students in a room who were unfamiliar with why that might be a problem—that's when the pain level increases to the point that I'm thinking, 'Boy, we need to be doing a better job.' Because as soon as we become comfortable with it, even without our being aware of it, you've migrated to— Now again, I'm a little nervous about talking to you about this, simply because it sounds like, and my colleagues here would say, 'Geez, you know, that doesn't make sense. I think this university does a very good job of trying to deal with that issue.' I think having WARF at arm's length from the university helps. They've been a tremendous success. Technology transfer has been extremely successful, not just in terms of benefits to the university but benefits worldwide. And we have to continue to do that. But I always want to be sure we do it with a conscience."

I asked Mulcahy whether the university permits confidentiality provisions in agreements with firms that provide support for research on campus.

"In our grants and contracts that come in with confidentiality agreements, we typically either work to change the language or we flat-out right don't do it. Now, granted, if a company is providing proprietary material or things like that, we would agree, I believe—it's not my department that does it—but we would agree to respect their proprietary interests there. But we aren't going to conduct sponsored research by them which prohibits us from being able to publish, to share it with students, et cetera, et cetera. So we've been pretty adamant about holding the line there," he said.

"I understand you will tolerate a brief delay in publication. Ninety days or something like that."

"Yeah, ninety would be the most. I've seen typically thirty to sixty days. And that would be to allow the sponsors of the research to review what's there in terms of proprietary or patentable information that would generate from their funding."

"Do you try to tutor your graduate students and scientists and postdocs and others in dealing with many of these issues that you've expressed concern about?"

"Right," he replied. "I think the universities as a whole are trying to do a much better job in that. I think they recognize—I hope they recognize—that if we don't provide that kind of training, then you end up someplace you don't want to be without knowing it. And so we are, through various aspects of responsible conduct-of-research training. One of the things that we're doing that helps in part of our conflict-of-interest management system is to have disclosure to the students when faculty have certain arrangements that the Conflict of Interest Committee judges as representing a moderate to high risk for potential conflict. We have them disclose to their students and postdocs and trainees that this relationship exists and inform them that they have the right to seek help if they believe at any point that that relationship is somehow hindering their satisfactory progress to degree. Or if the research they're doing is not independent and meritorious on its own rights but represents essentially work for an outside interest."

"Have any graduate students or postdocs or others come forward and said they feel the rules are being violated?"

"We've had one or two students basically say, 'Is this an issue, is this something I have to be concerned about?' But we have not had anybody step forward and say there's a problem here. This is a fairly new development, I'd say in the last two years or so."

"Considering that you've got thousands of people involved, and hundreds if not thousands of projects, then either people feel it's perilous to come forward or there's nothing to come forward about."

"I don't think anyone feels that it's perilous to come forward," Mulcahy responded, "because we've made it pretty clear that there are multiple channels where this can happen. So, I don't think that's an issue. But we've not had anybody come forward, and I think it's because in general the management conditions that have been put in place have been adequate in addressing the concerns of each of those constituencies, the students, the technicians in the lab, and things like that. But I would guess that at some point, someone's going to come forward. I'd be surprised if they didn't. I'd like to think that we have a pretty effective system in place. Some of our investigators might argue that what we put in place they might view as being a little too austere, but we feel that it's quite appropriate. We have a committee that is composed of faculty and staff with experience in these areas, who pass judgment independently on these cases. That helps a lot. It's a peer-review, if you will, process, not a bunch of administrators who are doing it. So, I think there's a lot of buy-in. I have to admit, there's some resistance, as well. We have faculty who aren't happy with some of our policy decisions."

Objections to Rules

"They feel they're too restrictive?" I asked.

"Right. Particularly in the clinical arena. Just recently we developed a clinical conflict-of-interest policy which is definitely less tolerant than our nonclinical conflict of interest. And that's predicated on the fact that we believe that where human subjects are involved, it deserves particular scrutiny, close scrutiny, and so we have lower thresholds for reporting, lower thresholds for what we would consider a management condition. And in some cases we basically have informed investigators that they can't serve as PIs on certain clinical trials as a result of that. So that's new in the culture, and there's always going to be some knee-jerk reaction to that. When we talk to people about why we're doing it, that usually resolves some of the problem. Some of the problem is resolved when we inform people that the situation that we find them in is discretionary. That they can make choices that determine what the outcome is. If they're really feeling very strongly that they want to do the clinical research, then perhaps they need to alter the relationship. So, when it's presented that way, some people change their relationship or they decide, 'Okay, I'll abide by these rules and find other ways to do it.'"

"What would be the criteria for excluding a PI from a particular project?" I asked.

"I would just use this as the textbook case, because clearly there are so many unique circumstances that deserve additional consideration, it's hard to pick a typical. But let's say a faculty member spun off a company to market a technology that they developed and the success in a clinical trial is obviously going to increase the value of that technology. We feel very strongly that they should not be in a position to run that trial, to evaluate that trial, to assign patients to that trial, so we basically say, 'You can't do it.'"

"Have some tried to do that?"

"They haven't tried to do it," Mulcahy said. "In other words, they haven't tried to force the issue. They may say, 'I don't understand.' It's very common to say, 'You're being too restrictive. This is not influencing how I would do this, that, or the other thing.' They may come forward and say, 'This trial is part of a national study. There's an independent data-monitoring board. I have very little direct influence on the outcome.' There are a lot of ways they can try to discuss it. But we've been pretty firm in saying unless you can identify compelling circumstances, we've held the ground. They can appeal to the Conflict of Interest Committee. It's the same committee, but in the case of an appeal, the PI has a much more direct and active role of presenting their own case against the decision. So, in a way, it hears an appeal of its own decision in some cases. And we thought that's very important because quite honestly, conflict-of-interest situations can get so sticky that if you don't have experience with them, it's easy to make a decision that doesn't make sense to someone who's got more experience. So, we think that's the most appropriate group to do it. The other thing we will do, and this is a process that's still in evolution, but we share the conflict-of-interest information with the Human Subjects Committee, and they are free obviously to impose additional conditions on certain types of research that they feel are necessary to ethically or morally protect the patients in the study, being more familiar with what the details of the study are than the Conflict of Interest Committee. There's no overlapping membership from those two committees, but when we developed the conflict-of-interest in clinical research policy,

I deliberately established a committee which was a fusion of the two. I had representatives from the IRB and the Conflict of Interest Committee. They worked to develop general principles, and from the principles, policies and procedures. So, there we had joint representation. We have clinicians on the Conflict of Interest Committee, but currently none of them are on the IRBs. They have in the past been on the IRBs, which has been invaluable to us."

"It strikes me that you've developed a rather odd situation," I said. "On the one hand, there's the push for commercialization. There's the feeling, particularly among politicians and maybe in the public, there's gold in them there laboratories, and these scientists are perhaps ignoring it, not doing all they could for society. On the other hand, you've put up a great web of defenses to make sure that ethical standards are observed and that commercialization doesn't go hog wild and override scientific values. Where is all this going?"

"Well, hopefully," Mulcahy replied, "the way I view this, a continued awareness of that pain that I was telling you about, and our being aware of the tension that exists and ever mindful of our need and responsibility to deal with it appropriately. In my opinion, there's nothing more precious or vital to the success of science than the integrity of those who practice science. So in my opinion, if we have to build high walls to preserve that integrity, it's our responsibility to do it. It's also our responsibility to get the gold out of them there hills. And, in fact, I would say in my experience, the criticism that there's gold in them there hills and they're ignoring it is not true. I think the present-day university is pretty well tuned in to the gold in them there hills, for all sorts of reasons. Does everything get translated? Certainly not, but I think the awareness of the potential is more acute now than it's ever been."

"WARF goes back eighty years," I noted.

"Right. And on this campus, WARF is recognized as an incredibly successful story, and I think it's been vital to the history and the success of the university. I think our faculty recognize WARF as a significant benefit to them, both in terms of what WARF returns to the university to support research, but also in the kind of advice that WARF can provide them, how WARF handles their technology. So, it's a big, big plus for the university. But again, we have to be sure that we have the right guidance system for the whole process so that we don't inadvertently end up doing things that we had not—that we had never intended, to begin with. It's a very healthy tension."

I noted that "Derek Bok seems to think that science is going the way of NCAA athletics."

"Again, I think that's an exaggeration of the case," Mulcahy countered. "But I would say this: the potential is there if you're not paying attention to it. I would like to think nothing is going to go the way of university athletics. But I do think there are areas where similar concerns could be cited. But what I like to say is, at a university like this, we try to be very attentive to prevent that from happening. Time will tell about our successes, but to me the take-home message is that as soon as we are comfortable with the relationship, I think we've got problems. I think we always have to be concerned about that relationship. We need to encourage it, but we have to be sure that we're paying attention to it."

Relations with Big Pharma

"The medical sector presents special problems," I suggested. "You've got the pharmaceutical industry constantly pressing against the borders."

"True," he agreed. "The relationship between universities and the pharmaceutical companies is critical to getting new treatments into the marketplace and available to patients. I do think that sometimes that relationship has been too cozy and not necessarily at institutional levels but certainly at individual levels. I think some of the reactions that we get to conflict-of-interest management plans in this area certainly attest to how comfortable the relationship has become in some cases. But I think we're trying to be sure that that doesn't again influence the kind of research that's done, the kind of patients that end up in studies. And it's challenging, because the pharmaceutical industry provides all sorts of support to the university. Not just direct support for research. They provide gifts to individual investigators; they support training programs on campus; they support important seminar series, which bring a lot of valuable information to our community. But you got to be sure that at the same time the influence isn't undue."

"What sort of safeguards do you put in place? The pharmaceutical industry has traditionally been very insidious at getting into academic settings."

"I think many times our own faculty are not aware of some of those connections when they get involved with pharmaceutical companies. So that has been a challenge to sort out, and that's where having members from the clinical faculty serving on some of these committees is so valuable, because they're far more attuned to those issues than those of us who are from the basic science side would be. There are a lot of opportunities there, and I think a lot of very positive things that come out of pharmaceutical involvement, but, again, I think you have to constantly be aware and watching for signs and putting up roadblocks where it's appropriate."

13 The Journals Revolt

Medical journals occupy a strategic junction in science. They publish the research findings that are ultimately embodied in the drugs and medical devices that physicians prescribe for their patients. For all parties involved in this movement, the journals are crucially important. Publication in peer-reviewed journals is indispensable for career advancement in academic research and its medical wing. The scientific culture allows no other way. Reports of clinical trials in peer-reviewed journals heavily influence the sale of medical goods, because practicing physicians rely on journals to keep them informed of new diagnostic methods and treatments. These circumstances entice manufacturers to employ wiles, inducements, and influence to receive favorable reports in the clinical literature, particularly in the most influential and widely read medical journals. Among the thousands of clinical publications published worldwide, the most prominent are *JAMA*, the *New England Journal of Medicine*, the *Archives of Internal Medicine*, the *British Medical Journal*, and the *Lancet*. With large readerships, these journals possess unparalleled prestige among researchers in the health sciences and practicing physicians seeking to keep abreast of the latest findings. Careful editing and screening of submissions give them credibility and luster, leading to a self-sustaining chain reaction: they attract readers because they attract important, groundbreaking papers, and they get the papers because they have the readers.

For many decades the medical grapevine has buzzed with reports of cooked results appearing in important medical journals. It was only in the mid-1980s that a defensive editorial reaction set in, initially inspired by a rash of fraudulent research papers by individual scientists. Many journals strengthened their peer-review and editorial processes, requiring, among other defenses, assurances that all authors listed on a paper had actually participated in the research and were cognizant of the contents. While the fraud problem stubbornly lingers on, attention has also turned to a separate issue of integrity in publication: commercial exploitation of the medical literature by mercenary authors. Some journals began to demand disclosure of financial connections between authors and the drugs and medical devices on which they reported. Authors were required to certify that they had indeed participated in the research and had not merely permitted use of their names on ghost-written papers supplied by drug manufacturers. Journals also applied pressure for drug manufacturers to make available all clinical-trial results—not just those favorable to their products. Other measures aimed at reform followed.

In all of these efforts to safeguard the medical literature, a leading figure was an English-born and -trained physician, Drummond Rennie, who moved to the United States in 1967 and joined the faculty of medicine at the University of Illinois at Chicago. Rennie served as deputy editor of the *New England Journal of Medicine* from 1977 to 1981 and has since been a deputy editor of *JAMA*, as well as an adjunct professor of medicine at the Institute for Health Policy Studies at the University of California, San Francisco. Dismayed by fraudulent papers and the commercial misuse of the scientific literature, Rennie in the late 1980s organized the International Congress on Peer Review and Biomedical Publication, which has evolved into a major force against contamination of the literature. Meeting periodically, the organization drew 470 participants from thirty-eight countries to its fifth meeting, in 2005. An old acquaintance, Rennie chatted with me by telephone on January 14, 2006.

I asked Rennie to assess progress in protecting the medical literature.

"What we're talking about," he began, "is the influence of money on research that my journal and other journals publish. The distorting influence of it. And this distorting influence is huge. This isn't just an opinion. This isn't a guess on my part. This is a measurable fact. So my view is based on a great deal of evidence in the form of papers where it's been measured and so on, measurements I've made myself," he said emphatically, referring to surveys that have found that academics in the pay of pharmaceutical firms tend to report favorably on the products manufactured by their patrons. "And it's also based on a feeling of unsureness about what my journal is publishing."

"Time and Time Again Being Lied To"

"And I mean by that," Rennie continued, "that time and again we have a wonderful paper that has gone through review and I thought is great, but then I realize that I *don't* know whether I can trust any part of it, or which parts I can trust, which is the same as saying, 'Can I trust it at all?' And, of course, that's based on evidence too. This miserable, sickening evidence of being kicked in the testicles over misleading clinical studies. Time and time again being lied to. And that's, of course, happened to the *Annals of Internal Medicine,* the *New England Journal,* and *JAMA,* and a whole host of other journals, too, where later it's been revealed that the authors were lying and withholding and changing and so on, all to make money for the sponsors. And so it's changed the conversation, so the conversation has changed. And I'd say the fact is these things are happening and keep happening, and they're happening, it seems, with increasing frequency. The fact that these are happening suggests that we're going down to hell, going down the toilet. But the fact that I'm able to write about them, that other people are able to scream and yell gives me hope.

"Now, a good example of this is the matter of the registration of clinical trials, which I've been writing about and calling for and crying for and so on for years, years. I'm just taking this as an example to show the good things and the bad things that are happening. Because it has been obvious to me since 1989, when I was educated in this, the first Peer Review Congress, that if you based your treatment, in this particular case of cancer of the ovary, on the published studies, you thought that the drugs were pretty good. But," Rennie continued, "if you based them on all the evidence, namely, of all those studies that had been registered carefully at the start, you found there was no effect whatsoever. Which showed very clearly that there were a lot of trials that were being suppressed. To balance out the good news, there was a lot of bad news. And that to me was a startling piece of evidence that didn't particularly fit in with biases on my part either way. I was surprised. The bad news is that for me to have been surprised in 1989 is *ludicrous* in retrospect. And it just shows you what a pathetic judge I am of the world. But the fact is, also in retrospect, nobody else grasped the point at all in 1989, so I feel good about that," Rennie said.

"What happened next?" I asked.

"The fight then starts," he replied, "and sometime, I think in April 2000, Kay Dickersin, now moved to Hopkins [where she is director of the Center for Clinical Trials, Johns Hopkins Bloomberg School of Public Health], she and I and somebody else got a bunch of fairly high-level pharmaceutical company executives in the O'Hare Hilton. The second floor of that is the anteroom to hell. It's a place you fly into, hold a meeting, and fly out of the same day, and everyone is sleep-deprived and so on. And we argued strenuously all day to have *all* trials registered at inception. And they said it will be a cold day in hell before that happens. And I threatened them and said, with no evidence whatsoever, that 'we'll make you do it, we'll make you do it. The journals will make you do it.' Having every confidence that the journals wouldn't, because we're not just talking about my friends in the *New England Journal* or *Annals* or whatever. I'm talking about all those other journals out there that go along very nicely with the drug companies. And so that was that. And I had my little temper tantrum and jumped up and down, and that was that. Then in 2003 I said we've got to educate the editors here. It's clear that my colleagues don't understand—lots of people don't understand the problem. Indeed, I got a call from Marcia Angell [former editor of the *New England Journal of Medicine*], who you would think would understand the problem, and it was to say, 'Drummond, until now'—this was about 2003—'I thought you were wrong about it, that it was a nonproblem. But, of course,' as she said, 'I realize you are right.'

"It's basic," Rennie explained, "because if the existence of all the trials that were started were known about, you can then write a report that says, 'We've looked at all the evidence. This drug works magically well. On the other hand, that's on the basis of two trials that have been published that we know of. There are seventy-eight other ones that were started and have been buried, so you'd better be cautious.' So, I said, 'We're going to write an article.' So we did.[1] And I wrote it for my colleagues more than anyone else. I didn't care if anyone else read it, except editors. And we published it in *JAMA*, and it had a big effect," Rennie said.

"The bad news," he quickly added, "is that in 2003 people needed convincing it was a good idea. The good news is that it had an effect, and it had an effect for all these reasons. First of all, it was taken up and plagiarized, I may say, by the child psychiatrists, who brought it up in 2004 at the annual meeting of the AMA. I've never been to one and we have nothing to do with the AMA, as you well know. In fact, we cannot have, but they made a fuss. But my general feeling is that the only thing that makes any difference to anyone is a whole lot of kids dying. That gets congressional hearings, and a lot of pompous people pretending they were doing it all along. Fortunately, [New York attorney general] Eliot Spitzer brought his suit in June 2004 [against GlaxoSmithKline concerning the safety of the drug Paxil as an antidepressant for children]. And suddenly all the companies were saying they wanted to register clinical trials all along, or they did do it all along. And we'd already published a paper showing that they *didn't* do it and they *refused* to do it. And when they said they did it, they didn't. They perverted it by giving no information. Or they would give meaningless information. Where they were asked the name of drug, they'd reply, 'Investigational drug,' which tells you nothing. It destroys the whole system, but which, to a senator means, yeah, they're registering.

"So, all those things came together," Rennie said. "First of all, the stuff to do with these SSRIs [selective serotonin reuptake inhibitors], the question of suicide and all that in kids, and I loved Spitzer's approach, which was, 'I don't give a shit about the science. Anyone can understand that if you're hiding results, that can't be good.' And I love that approach. It's sort of brutal, effective, and it takes people less than thirty seconds to say 'Right on!' at a cocktail party. And I got suddenly a blizzard of e-mails from Eli Lilly, saying would I help them? They'd always wanted to register trials. Now would I help them, et cetera? And I said, 'Look, glad you're doing this and everything, but the answer is no. It would compromise my editorial integrity, and so on, and good luck.' Then comes the Vioxx stuff, and all the clear evidence that stuff has been hidden was coming out. That's the good side. And the people were relating this to actual harm, to real actual patients. People dying. And courageous people, like David Graham [of the FDA, who criticized the approval of Vioxx] and so on, were helping to make this happen. What's good about it," Rennie said, "is that it's coming out. And the excuse, then, that the companies were constantly making, that any form of regulation will inhibit production of drugs and so on, was just shown to be extraordinarily thin, self-serving, and bad. That's the good side. On the other hand, you've only got to hear a few politicians repeating it, word for word, from all those lobbyists, and that's bad. But I've done a study, which shows that the registration is zooming; we've had a wonderful effect. All the editors in chief got together and said we won't accept anything but trials that have been registered. Something that I hoped for in 1989, but it happened in 2004, and that's good. Very good. But it's difficult to get people to understand that you've got to do it at the start of the trial, not at the time of publication, because then it hasn't done anything. It's very difficult to understand that it's the *existence* of trials that you have to know about. So that as usual, there's a tremendous amount of wiggle room left for Pharma to creep out. And so I think the whole trial registration business encapsulates very well my feelings about pretty well everything else that's happened in the area of Pharma."

Deaths Make a Difference

"I'm not going to talk about the massive evidence for buying of doctors," Rennie said. "The idea you get the doctor hooked on prescribing the drug first and then they can't prescribe anything else. That's so obvious. I think it's just very good it comes out, and of course I hope that enough people read enough and think about it enough and get cynical enough about it. But it is obvious to me that the only thing that will really change things is some massive amount, some real series of deaths, which are directly related to something."

"Something like the Gelsinger case on a large scale?" I asked.

"Yes," Rennie replied. "Isn't it interesting. In the matter of the [calcium] blockers [for heart disease], they estimate more people were killed by the FDA permitting this drug—approval was fast—than were killed, twice as many as, in the Vietnam War. Vioxx has certainly killed a ton of people but may be quite a good drug, in my opinion. The scale of things is difficult for people to grasp, and I don't know how to get that across. So, I'd say it's very good that laypeople know so much about it. It depresses me that medical, clinical research scientists, my buddies, are so extraordinarily resistant to the idea that they can be and are influenced by money, like the rest of the world."

"It would seem," I said, "that the academic institutions, the medical centers, would be a strategic place for enforcing right behavior."

"I think so," Rennie agreed.

"But they're quite flabby," I added. "They don't do very much."

"Basically, the academic medical centers have to be pretty damn big and powerful and sure of themselves not to get engaged in a race to the ethical bottom. Because they keep thinking, 'If we don't do it, somebody else will get this money.' But the fact is they're much better now than they were, so that's a plus."

"Most universities do have conflict-of-interest regulations. Whether or not they enforce them is a separate matter," I said.

"First of all," Rennie replied, "there's the enforcement. Secondly, there's the regulation. UCSF has got some very good rules. But I think the AAMC rules are ridiculous. The idea that it's immaterial if you get ten thousand bucks—that's where the rules start kicking in. It's just fantastic. I can be bought for a lot less than— I can certainly be influenced for less."

"Do you think the $10,000 rule is ineffective?"

"I think it's silly, because it doesn't work on money. And I don't know how to make rules about this. I don't know how you make rules, given that we know it's the relationship, it's the gift relationship that matters. And the very fact that you've taken a gift means you can't help behaving differently toward the donor. You know that that's hard work. So that immediately flattens out the difference between a hundred bucks and ten thousand bucks. But the fact is ten thousand bucks means a different—a kid going to college and all that."

I asked, "Is it possible to maintain integrity and the purity of the literature and the scientific process and at the same time have these guys and a few women making a lot of money from Big Pharma?"

"No, no," Rennie replied immediately.

"There's no way to sanitize this?"

"There's no way. I'm talking about what we deal with in *JAMA*. We don't deal with the new molecules," meaning basic scientific research, remote from drugs. "And there, I don't think it matters too much. But let's say you've developed this brand-new drug that treats hypertension in a new way. I'm strongly in favor of those people making some money as a consequence, the university making money as a consequence. Developing it and all the chemistry and so on. I love the thought that they get money, as well. But the university is faced with the problem, of course, of how much time they're putting into it, and whether all the fellows are just slaves for this little corporation, or what, and how do you apportion time and energy and all that. Those can be dealt with, I believe. What I don't think can be dealt with is the *testing*. And that's where we—we're in that end, you know. The *New England Journal* and *JAMA*—our articles about drugs have to do with how do they work and what harm do they cause. And so most of them, certainly the early ones of any drugs or clinical trials of these, are all, to an astounding extent, under the control of the sponsors, who have the most to make from a positive result. And when I say under the control, you saw it beautifully in the case of Vioxx. In the Advantage Trial.*

"I mean, for god's sake," Rennie said, "it's all there in the record. So Merck pressured the doctors to change the cause of death. There's plenty of evidence of that in e-mails. They threatened the FDA if they highlighted the cardiac arrest. There's lots of evidence. They hid three extra Vioxx heart attacks. I mean, that's what's going on. We're well aware of it. We're trying to winkle it out. We demand that studies have independent statistical review. So, of course, you can see how much I care about trust. There's trust for you."

"The clinical trials are going abroad," I said. "That would seem in many ways to multiply the problems."

"Which of course is a way to multiply the problems," Rennie agreed. "We've got, for better or worse—very much for better actually—we've got a system for dealing with scientific misconduct here, fabrication, plagiarism. And it may be lousy, it may be this, it may be that, but in general now, scientists have reached the absolutely staggering conclusion—it's generally accepted—that research can be cooked and faked. And so you've got a system for dealing with it, and the general heat has gone down, and it's automatic, and you don't usually have it as front-page news. But we've got this remarkable thing happening that if you get problems with a paper from abroad, you can forget it. So that there are two standards coming in. A paper coming from Italy and a paper coming from Johns Hopkins. On one, there's forced accountability. In Italy, you can just whistle to the pope as far as we're concerned. Nothing's going to happen."

"There is no way to find out what's going on?" I asked.

---

*A clinical trial sponsored by Merck, comparing Merck's Vioxx and a competing drug, naproxen, for the treatment of osteoarthritis. Merck was later accused of not reporting the death of a woman, among several other fatalities, in the Vioxx trial and withdrew the drug in September 2005. The drug trial was published in *Annals of Internal Medicine*, October 7, 2003, and details concerning the publication were reported in the *New York Times* ("Evidence in Vioxx Suit Shows Intervention by Merck Officials," April 24, 2005). In the *Times* report, the lead author on the *Annals* article, Jeffrey R. Lisse of the University of Arizona, is quoted as saying: "Merck designed the trial, paid for the trial, ran the trial. . . . Merck came to me after the study was completed and said, 'We want your help to work on the paper.' The initial paper was written at Merck, and then it was sent to me for editing." The *Times* report added that "Dr. Lisse said he had never heard of the woman who died, until told of it by a reporter. 'Basically, I went with the cardiovascular data that was sent to me,' he said."

"The University of Pisa, or whatever it is, may pretend that they've got something. But first of all, they've got to define misconduct, and then they'll say, 'Well, you know, the guy is this or that; he's very important.' There's no one there. You can't do it. I only mention that because of trials going abroad means you actually have no controls. What sort of controls do they have about anything there? I don't know," he said despairingly.

"Is any pressure brought to bear on *JAMA* in response to what you've been writing over these many years?"

"No, not really," Rennie said. "People get very angry with me personally. Very angry. Won't speak with me. The trouble is when you write about this stuff, you feel like a pompous jackass. At least I do. Because it sounds as though I'm preaching. And I don't want to preach." Referring to his avocation, he said, "On the other hand, if you're a mountain climber, you actually believe in trust. Roping up with somebody who's going to save your life matters. Well, then, you realize trust matters. And then you think, the whole of science depends on trust. You can't have it set up like a bank-checking system that everyone— You just can't do it.

"Going back to the climbing," Rennie said. "I'm a really bad climber, a third-rate climber. But I've been on a lot of expeditions, for this reason or that. I love it and have usually climbed with people who have been terrific climbers and did nothing else. But they're buddies of mine, and all that. But especially if you're a third-rate climber, you look on the rope as a metaphor for trust and you care about trust. It all comes down to that, and I don't know how we can sustain a system where everyone is busy losing trust."

Quality Control Uncertain

"You receive a paper based on a large multicenter trial," I said, "a good deal of which was conducted abroad, in eastern Europe, India, Asia. What assurance do you have of quality control?"

"I don't know. That's it. I don't know," Rennie replied.

"So a journal has to put forth a great deal of trust," I said.

"We have to, yes," Rennie answered. "And it's the same when we publish an ad. We can say that's disgusting, we're not going to publish it. I suppose there must be people at *JAMA* who do that. But going into all the claims and so on, god, that would be a whole-time job for all the editors. So in the end we put a lot of trust there. And in clinical trials, usually it comes down to the small committee at the top who thought up the trial and so on. And we hold them responsible, and we ask them a lot about what they did and how they did it and all that. But, god, if they're lying in their teeth, we've had it."

"You would not invest much trust or faith in Big Pharma?" I asked.

"We don't. And in fact, it's worse than that. Because often the most important papers, and the best papers, come from them. And so we're in an agony about that. But sometimes we turn them down, because we simply don't believe that we can know enough about the papers. And, of course, as I've said already, the Advantage study and all that, all published by trusting editors—because you can't know. How could we have known, if they say this is what we did and if they sign a form saying they've given us all the data and everything, but they didn't—they knew that they didn't—and so on? What can we do, except tell the world?"

"I get the impression," I said, "that pharmaceutical firms have been particularly manipulative with the journals to promote off-label use [of drugs for purposes other than those approved by the FDA]."

"Yes," Rennie agreed, "they certainly are, and this is very troublesome to us, because we think the laws are ludicrous. But you're often stuck with that in the end. It's sort of pathetic how much—I have to use the word 'pathetic.' But here we have all this stuff, right in front of us, these data, you read it, and of course, whatever it is, is any of it true? You get it reviewed. Sometimes the reviewer saying, 'It's a lovely paper; I wonder if it's true.' And then you say, 'Well, we need an independent statistical analysis,' and deliver two truckloads of stuff to such-and-such a statistician, and in three months' time we will get an answer. Then he says it's okay, maybe needs a few changes and so on. And then we think, 'I suppose it's okay.' The problem with it is that you can't then change it. It's very difficult to change a paper so much that it blocks anyone from using it for any off-label use. If the facts are that, yes, this cardiac drug does help a little bit with acne—one of the things that was significant. You can say, 'But this is just a subset.' But in no time at all, you see it's very difficult to control the use made of a paper. Even if you change the conclusion from 'this works brilliantly' to 'it's possible that this works.' Stuff like that, which we do all the time. But you can't really control things after that time," Rennie said.

"The disclosures that you now require about financial connections to the content of a paper and things of that sort, is that just simply inducing complacency on the part of the reader?" I asked.

"First of all, I think it's terribly important you do it," Rennie replied. "But, of course, you're an idiot if you think that solves the problem.

This is what happens: You have an entire page in small print disclosing financial conflicts. But you don't know which of them matters. You see 'Drummond Rennie gets money from the A, B, and C.' From A, I got a ticket from Medford, Oregon, to San Francisco to attend a conference, a panel discussion. From B, I got two and a half million bucks and god knows how many shares. But, you see, it all looks the same."

"Can you have them specify how much from whom?"

"We demand everything. You specify as much as you can. But we don't have the sort of manpower to go into Standard & Poor's or Dow Jones to look up who's director and so on. Because that's a huge thing to do, and because people lie to us *all the time*. And then they claim they didn't understand the meaning of the term 'money' or 'consult' or whatever. We've had that. 'Well, we thought you meant a lot of money.'" With a laugh, Rennie added, "Who's to say two million is a lot?

"I take the view," he said, "that universities, everybody, would benefit if trials were relevant, first of all relevant to patients. Secondly, were believable. Everybody would—the researchers, the journals, the universities, above all, the drug companies. How do you make them believable? They can do their own studies, and they can publish their own studies as much as they like. But on top of that, you need to have a national center for doing studies, which says we're going to actually do a study that matters, such as the AHRQ [U.S. Agency for Healthcare Research and Quality, which conducts research on medical treatments]. There was a study on a gigantic number of patients which we published a few years ago, which showed that the cheapest medicines, the oldest and cheapest medicines for hypertension did a better job—had less deaths, had better control—than the newer ones.[2] And the older drugs did it at one-hundredth or one-thousandth the cost. And of course it came under huge siege. The researchers were said to be too old, they were too young, they were too experienced, they weren't experienced enough, they were this, they were that. You know, ad hominem attacks and so on. That's the sort of study where we can say, 'What are the major issues we've got to solve?' They didn't ask, 'Do we need another five diuretics?' We've got 250 on the market already. It is, how do you get a better pain pill the elderly folk can take for the rest of their life? So you do a long study, not Vioxx for three days, since they're going to be taking it for life. And they contract out to universities, and no one's allowed—it all goes through universities, or whatever. No one's allowed to take any penny that isn't like an NIH grant. And the results would be enormously beneficial, or not, to companies, because they could then say, 'Our drug actually works.' Of course, they don't want to do it, because it's never a brand-new drug, but it might be. But people would believe it, and of course then we'd be on safer ground. And it's got to be done, and it's never going to happen under the Bush administration or anything like it."

Part Three: Fixing the System

14 What's Right and Wrong, and How to Make It Better

The ivory tower is papered with contracts, patents, and business plans, and the pathways to academic laboratories, if not paved with gold, are strewn with stock options—and ethical pitfalls. So it may seem from the polemical din around our subject. The critics of commercialization hold that in unrestrained pursuit of money, academic research has gone ethically numb, to the detriment of science and the public well-being. The promoters, practitioners, and admirers of scientific entrepreneurship ridicule and reject that judgment. They claim that important societal benefits come from university-based business pursuits, while conceding, sometimes, that ethical failings occasionally occur and need correction.

Both assessments depend on where and when you look. But be careful. Academic science is a huge, diverse, and mutating enterprise. The penetration of commercialism varies among universities, from pervasive to slight, and the observance and enforcement of ethical standards and regulations range from strict to sloppy. The same applies to important offshoots of academic research, particularly the major scientific and medical journals that, justifiably or not, put a stamp of validity on the research they publish. Partisans favorable to business on campus downplay episodes of inexcusable misbehavior as rare aberrations. Emphasizing, or exaggerating, the financial and social benefits of academic-corporate collaborations, they cite steps taken in recent years to protect and elevate scientific integrity. These backers tend to be Pollyannaish, glossing over the difficulties created by academic-business dealings, while dwelling upon the dollar returns, jobs created, and products delivered to the market. The critics of science for sale express revulsion at ethical misdeeds committed under the banner of science. Long before the Bayh-Dole Act, they argue, academe contributed its scientific knowledge to society without the entanglements of personal and institutional enrichment and the infiltration of sharp business practices. They are particularly incensed by the whoring of academic research for pharmaceutical money. Unless bound by strict safeguards, they insist, entrepreneurship is a menace to academic values and public safety. Even with strict safeguards, some contend, it deserves no place in academe. The critics argue that the vaunted economic benefits are inflated, even illusory, that the damage inflicted on the scientific commons entails societal costs that are not acknowledged in the triumphant bookkeeping of Bayh-Dole. Both camps can be derisive and extravagant in their rhetorical exchanges.

Much of this contention is ideologically rigid, indifferent to the upgrades in ethical sensitivity and enforcement that have occurred in academe in recent years, but also indifferent to the stubborn persistence of unethical behavior in contemporary science, sometimes brazen, sometimes stealthy, but seemingly ineradicable. Overall, for protecting the integrity of science and reaping its benefits for society, wholesome developments now outweigh egregious failings—though not by a wide margin. Nonetheless, the changes and trends are hopeful.

That's my conclusion after talking to people in universities, medical centers, and elsewhere who are knowledgeable about the commercialization of science. As I observed at the outset, shame and embarrassment exercise great force in academic and scientific affairs. Pride plays a big role, too. Scientists, their managers, and their institutions normally care deeply about their reputations. When they go wrong or miscalculate, the odds for exposure, harmful publicity, and unpleasant consequences have greatly increased. Prestige and funds may be lost; careers may be harmed. In the last resort, they may be confronted by government authority, as we saw with the federally ordered shutdowns at Duke, Hopkins, and elsewhere. Their humiliation reverberated throughout the research community. In this period of anti-government government, typified by the supine Food and Drug Administration, regulatory authority is limp and deferential to business pursuits, at the price of scientific integrity and public safety. Even so, the scientific norms of honest dealing create a vulnerability to the judgments of colleagues and the public, including the donors who are tirelessly courted by the mendicants of higher education and science. This vulnerability has contributed to beneficial changes in the ethical sensitivity and behavior of the scientific enterprise and the taming of reckless, negligent, and overtly greedy tactics. However, there's still some distance to go. And there's always the danger of ethical backsliding by individuals and institutions that surely know better. As numerous episodes confirm, humankind's plentiful capacity for foolish to abominable behavior is not neutralized by advanced degrees or high academic position.

Lessons from Confronting Scientific Fraud

We expect academic science to behave better than the society in which it is embedded. The high standards of truthfulness and ethical compliance that are historically engraved into the scientific culture have no counterpart in business, government, or other sectors of society. Scandals on Wall Street, in Congress, and in the clergy have become ho-hum. Not so in science, where achievements and transgressions draw wide public attention. Report a breakthrough, and it's big news; be found to have faked it, and that's big news, too. That sequence has been played out many times, from the Piltdown hoax to the cloning hoax in South Korea. "The truth, the whole truth, and nothing but the truth" comes from the law but bespeaks the ideals of science. That the high standards in science are sometimes violated is obvious. But concern about scientific honesty is abundant because misdeeds are widely regarded, by scientists and the public, as intolerable, dangerous, aberrant. The violations, when discovered, are almost always deplored and punished. The record provides reason for hope for cleansing the relationship between science and moneymaking.

This expectation on my part borrows from the considerable progress in recent years in strengthening a major requirement of ethical behavior in science—the honest conduct and reporting of laboratory research. There's some overlap here with science in pursuit of commercial gain. But in the categories of scientific deviance, faked experiments and doctored reports are in a separate domain, usually inspired by the pursuit of professional recognition, tenure, or glory, rather then money. The perpetrators work alone, or at most with a colleague or two, and though their mentors and institutions may smile on their faked accomplishments prior to exposure, once revealed they face professional doom. Such offenses against truthfulness are officially branded "scientific misconduct"—fraud, in common parlance. As a violation of federal regulations when government research money is involved, they come under the authority of the U.S. Office of Research Integrity, which is specifically concerned with a narrowly defined band of offenses: "fabrication, falsification, or plagiarism in proposing, performing, or reviewing research, or in reporting research results."

ORI is not concerned with the greater complexities that arise from academic commercialism, such as financial conflicts of interest, professorial shilling for pharmaceutical products, concealment of business connections, and shortcuts in protecting patients in medical experiments. In the division of jurisdictions among federal agencies, some of those issues fall to the Office for Protection from Research Risks, but most float free in an unregulated zone between academe and commerce. In its sector, ORI has achieved impressive progress. Fabrication, falsification, and plagiarism are still with us, as are the other sins in ORI's mandate. But when it comes to the honest conduct and reporting of laboratory research, science today is more like an orderly, well-policed metropolis rather than the ungoverned territory that it once resembled. From this, draw hope but recognize that time and great effort were required to achieve the result.

Allegations of scientific fraud can be difficult and painful to investigate. Whistle-blowers can be correct, mistaken, or purely malicious in making accusations. In some instances, they are graduate students, postdocs, or lab technicians, low in the scientific hierarchy, vulnerable to retaliation, and unsteady in testifying. The unseemly tag "disgruntled" is often applied to those who disturb scientific peace. Fraud cases often entail murky circumstances compounded by personal animosities. They threaten professional reputations and careers with defamation and can burden universities with expensive litigation and unwelcome publicity. Research institutions thus had many incentives to avoid confrontation with fraud. The claim that the fraud issue was itself fraudulent lingered long among the leaders of science, even after a series of egregious misdeeds were placed on the public record. Public fascination with fakery in science evoked assurances that the working methods of research reduced fakery to near nonexistence—so that "99.9999 percent of [scientific] reports are accurate and truthful," the editor of *Science,* Daniel E. Koshland, editorialized in 1987.[1]

The reasoning behind this confidence, though unsupported by evidence, was superficially plausible and ran as follows: Scientific results, faked or authentic, must be published for the investigator to receive credit, thus ensuring scrutiny, attempts at replication by other scientists, and, sooner or later, exposure of fakery. Deterrence was provided by the impossibility of a clean getaway. To the trivial extent that fraud did occur, the apologia proceeded, it was attributable to a few slippery sociopaths in white coats, bogus members of the profession, destined to be caught, expelled, and their tainted reports expunged from the scientific literature.

The fraud issue was dismissed as a creation of overheated journalists and opportunistic politicians, including a young congressman, Al Gore (D-Tennessee), who in 1981 chaired the first congressional hearings on scientific misconduct. Based on several cases that had come to public attention, including the infamous fakery of successful tissue transplants at the Sloan-Kettering Institute for Cancer Research (see p. 36), the hearings brought scientific misconduct to public attention. But the problem was still depicted by the scientific establishment as small, under control, and inconsequential to the progress of science.

Between 1974 and 1981, merely twelve cases of scientific misconduct were disclosed in the United States, according to a historical review by ORI.[2] During those years American scientists, in all fields of research, published a total of nearly one million papers.[3] The minuscule count of fraudulent papers, though improbably small, seemed to substantiate the self-cleansing theory of scientific rectitude. In the absence of a crisis or embarrassingly rampant scandal, the buck was passed for responsibility for keeping science honest. Universities contended that research journals should screen submissions for fraud, while journal editors pleaded lack of resources and urged academe to take up the task, while all agreed that the issue did not warrant great concern. However, starting in the 1970s, erosion of trust became a prominent factor in relations between science and society. The Tuskegee scandal led to the passage in 1974 of the National Research Act, which required universities to establish review boards to monitor the safety of volunteers in clinical research. There was no counterpart legislation focused on fraud in the laboratory, but that issue, too, attracted scrutiny and public attention.

Allegations of scientific misconduct as a stain on modern science gained credibility with the 1982 publication of a book providing case-by-case accounts, *Betrayers of the Truth: Fraud and Deceit in the Halls of Science* (Simon and Schuster), by *New York Times* reporters William Broad and Nicholas Wade. In 1985 the Health Research Extension Act directed the Department of Health and Human Services to require institutional recipients of its research money to establish procedures for reviewing and reporting to the HHS incidents of "alleged scientific fraud" in research supported by the NIH and other parts of the Public Health Service. Scientists, the press, and Congress moved to a higher state of ethical alert. Within the scientific leadership, present, too, were the old fears we've encountered that loss of the scientific halo would lead to loss of money. Cases that were once dealt with quietly or ignored now received official attention and came into public view. A pair of maverick scientists at the National Institutes of Health, Ned Feder and Walter Stewart, went public with allegations of official indifference to scientific misdeeds, leading to a series of congressional hearings and numerous reports of scientific misconduct in the popular press.

The dishonor roll included several young, fast-rising university-based researchers who published papers at a phenomenal pace—sometimes as often as one a week over long stretches of time—with their approving professor-mentors flattered to have their own names tacked onto the papers in a custom known as "honorary authorship." When these Stakhanovite feats of productivity were found to be based on deceit rather than discovery, the mentors scurried away, claiming innocent involvement, even unawareness of the presence of their names on papers published in prominent journals.

Starting in 1986, Nobelist David Baltimore's stout defense of a co-author accused of misconduct played out intermittently on Capitol Hill and in the press for nearly a decade until an administrative court in the Department of Health and Human Services finally exonerated the accused. Nonetheless, the protracted proceedings had generated numerous press reports that blurred together science, fraud accusations, Nobel prize winner. Though Baltimore was not accused of any wrongdoing, the proceedings became known as "the Baltimore case." Appointed to the presidency of Rockefeller University in 1990, Baltimore could not shake off the stigma of the case and resigned under pressure in 1991, after eighteen months in office. Scientists at Rockefeller feared that his presence would put them at a disadvantage in competing for grants, it was widely reported. Taint by association was difficult to dispel, even for a distinguished scientist. Six years later, with passions over the case having cooled, Baltimore was appointed president of Caltech, where he served successfully for a decade before retiring.

Press reports and congressional investigations sullied the good name of science to the point where the chieftains of the profession finally saw the need to appear attentive to the fraud issue—though they considered it greatly overblown. A federal agency for assuring honesty in research, which later was named the Office of Research Integrity, was established in 1989 within the NIH. At the National Science Foundation, the policing role was assigned to the Office of Inspector General. Under federal regulations and prodding over the past fifteen years or so, rules and procedures have evolved for promptly investigating allegations of such scientific misdeeds.

At every university receiving federal research funds, procedures are in place—or are supposed to be—to investigate allegations of fabrication, falsification, and plagiarism, and to protect whistle-blowers. The prescribed procedure calls for securing laboratory records when allegations of misconduct are made—an important step, often neglected in the head-in-the-sand era. Instruction in ethical scientific behavior is required for all research staff supported by federal money. Mindful of their reputations, universities usually now respond to allegations of misconduct quickly and openly. When flaws are detected in published papers, authors often hasten to explain or withdraw them. The system has its failings. Some researchers may succeed with an episode or two of sleazy behavior, and now and then you hear about someone who has managed a full career's worth.

Nonetheless, the overarching reality is that the research community now recognizes that a substantiated allegation of misconduct is lethal to a scientific career and harmful to an institution's reputation. In the past, the federal government limited official punishment to short-term or permanent exclusion from grant eligibility, thus branding offenders as untouchable and effectively drumming them out of science. But now, even harsher consequences have ensued. In 2006, for the first time ever, according to federal prosecutors, a finding of scientific misconduct resulted in a criminal prosecution and imprisonment for making false statements on grant applications. In this instance, a one-year prison sentence was imposed on a prizewinning medical researcher, Professor Eric Poehlman, formerly at the University of Vermont College of Medicine, who also repaid $180,000 to the government and retracted ten published papers.[4] But just as centuries of hanging have not abolished murder, detection and punishment will never abolish scientific fraud. As a surreptitious offense, its incidence defies exactitude. However, there's no evidence of a flood of misdeeds in laboratory research. Rather, within the NIH and NSF jurisdictions, there is a relatively small but steady flow of allegations of misconduct, but upon inquiry at the local level, few are officially determined to rate the damning brand of scientific misconduct.

For the years 1992–2001, the Office of Research Integrity reported that 248 institutions reported 833 allegations of misconduct, resulting in 110 findings of misconduct. The number of misconduct investigations initiated by institutions totaled 12 in 2003, 27 in 2004, and 21 in 2005. Findings of misconduct in these cases totaled 12 in 2003, 8 in 2004, and 8 in 2005.[5] Skeptics scoff that rampant fraud goes undetected or tolerated in science. That's the claim of Horace Freeland Judson, an honored science writer, in *The Great Betrayal: Fraud in Science* (Harcourt, 2004)—a work long on assertion but short on evidence to back the author's claim of epidemic fraud. Judson may be correct, but valid allegations of scandal usually ignite a follow-up rush of investigations and findings, journalistic, scholarly, and political. So far, that has not occurred.

The globalization of science has internationalized sensitivity to scientific misconduct. In one of the most notorious fraud cases of recent times, the false claims of human-cell cloning by the Korean scientist Hwang Woo Suk brought his swift dismissal from Seoul National University, retraction of his published papers in the American journal *Science,* and apologies and public introspection at that journal about improvement in its screening methods for publication. Hwang's American collaborator—Gerald P. Schatten, of the University of Pittsburgh—was promptly brought before an inquiry convened by the university, strongly criticized, and deemed guilty of "research misbehavior."[6] No penalty was attached to that finding, which is not in the federal lexicon of scientific wrongdoing. In any case, the University of Pittsburgh was visibly anxious to demonstrate publicly its commitment to upholding scientific integrity. The outcome of the case was not career-enhancing.

Trust is often cited as the social cement of science. But trust alone has never sufficed. Safeguards against misdeeds are historically built into science. In today's circumstances, they are more important than ever and merit strengthening. "How do we tell that particular scientists are speaking the truth about the world?" asks Steven Shapin, a historian of science, in *A Social History of Truth: Civility and Science in Seventeenth-Century England:*

> We inspect them for the recognized insignia of affiliation with the institutions in which expertise lives. How do we tell that the institutions harbor genuine knowledge? We inspect them for signs of internal "rigorous policing" or are otherwise assured that institutional control has been exerted against the passions and interests of their members. Who would not misrepresent the truth for advantage if they could get away with it?[7]

The task ahead is to minimize the negative effects of science for sale in activities so far resistant to or only slowly adopting ethical safeguards. These are mainly in clinical research, areas in which commercial money and academic interest and needs intersect, and independent scrutiny is resisted or refused. We can take heart from success against the evil trio under ORI's surveillance: fabrication, falsification, and plagiarism. They have not been abolished but have been drastically minimized by strict regulation, ostracism of offenders, and, recently, imprisonment.

Focused on three categories of clearly defined behavior that have long been abhorred, the crackdown on scientific misconduct was relatively easy to muster and enforce. The sins arising from scientific commercialism pose a far more challenging problem: keeping science honest while potent forces push it hard to make money.

A Troubled but Durable Relationship

The productivity of contemporary science is evident in the proliferation of healing drugs, devices, and methods derived from both incremental and revolutionary growth in biological understanding. Yes, academic-industrial collaboration should be more closely focused on societal needs and benefits, and less on profit-seeking. The billions spent by Big Pharma on developing and marketing copycat versions of competitors' drugs are a testimonial to corporate irresponsibility. In addition, the pharmaceutical industry's unwavering pursuit of extravagant profits and its crafty, sometimes dangerous, marketing strategies radiate a baleful influence on the life sciences, the medical profession, and public confidence. In a rare admission, a leading industry official acknowledged the negative publicity arising from corporate misdeeds. "I've been asked, 'How can we stop bad stories?'" Billy Tauzin, the former congressman who heads Big Pharma, remarked at an industry forum in 2005. Tauzin's answer: "Let's stop doing bad things."[8] The industry has done so many bad things that medical journals are increasingly wary of any submission directly or indirectly touched by its money or people, and an increasing number of medical schools restrict or prohibit pharmaceutical representatives from on-campus contacts with their students.

But even with misguided priorities and reprehensible behavior, the academic-industrial research system produces beneficial results. Equally important, the two sectors are cemented by politics. Reformers are at liberty to dream, rail, scold, and campaign, but any thought of prying academic science and business apart is detached from political reality in modern-day capitalist America. Big Pharma runs one of the biggest lobbying operations in Washington, and pharmaceutical interests are among the biggest providers of election-campaign money. Tauzin's position at the head of Big Pharma, with another former congressman heading its biotech counterpart, the Biotechnology Industry Organization, presents a lesson in post-congressional employment opportunity for legislators looking ahead from Capitol Hill.

With strong backing from many quarters, the relationship between academic science and business is firmly entrenched in the American knowledge system. It is strongly supported by academic leaders, somewhat less so by industry, but only because of corporate qualms about the difficulties of doing business with ethically sensitive academic science. With Washington paying the majority of the costs of research, industry is keen for a thriving academic science sector—at little or no cost to itself, but with ample opportunities to cherry-pick the fruits of academic research and invoke the prestige of its superstars for promoting sales. The overall backing for the relationship is substantial and the opposition is weak. Where is the head of a research university who today urges extreme caution in scientific dealings with industry, or who declines any dealings at all as too laden with ethical hazards to be worth the risk of public endangerment, reputational harm, and opprobrium? I've never heard of one. Vows to increase connections with business and industry, to serve as an economic engine for the community and the nation, are boilerplate passages in the addresses of university chieftains. The intensity of academic-industrial relations varies from place to place, but, overall, the two sectors are either intimately linked or they're pawing each other in search of collaborations. If Congress threatens to reduce funding for university-based science, corporate leaders warn of damage to American industrial competitiveness. Corporate financing of academic science is piddling, but dollar-savvy CEOs know they need it, and they press the government to provide the money.

Universities produce knowledge, and industry manufactures and sells goods. On that basis, they do business, amid widespread applause and strong encouragement to do more. The academic holdouts from business relationships are few to nonexistent. Even Harvard, long reluctant to engage vigorously in tech-transfer pursuits, has mobilized to expand its share, while delegations from around the world seek instruction in academic entrepreneurship at American universities.

The ethical risks of doing business with industry are well understood in higher education. But the hunger for both money and the appearance of performing public service override caution. Revelatory nuggets are sometimes present in the arid prose of academic policy pronouncements. An illuminating find was contained in a cautionary statement recently published in *JAMA* in the cause of protecting medical research, training, and practice against the unsavory tactics of Big Pharma. Signed by eleven senior figures in those fields, the statement argued that the sample drugs, expensive meals, travel funds, trinkets, and other gratuities that pharmaceutical representatives routinely bestow on medical students and physicians can unconsciously warp clinical judgment and should be banished from academic medical centers (AMCs). The statement also deplored corporate influence over continuing medical education programs, ghostwritten papers to promote drugs, and pharmaceutical sponsorship of lucrative marketing pitches by mercenary scientists posing as objective researchers.

Similar pleadings for scientific and medical rectitude have been periodically sounded for many years. If they have any effect, it is never long-lasting, for not many years later, the same alarms and remedies are restated. Like their predecessors in appeals for ending these briberies, the *JAMA* signatories argued that the giveaways and subsidies are corrupting influences. The guidelines developed by various professional associations and Big Pharma "are not sufficiently stringent and do not adequately uphold a professional commitment to patient welfare and research integrity," the statement cautioned. It recommended that the academic medical centers themselves "must be prepared to monitor compliance and enforce the rules we have outlined." Amen. But though reflecting stark distrust of industry, the rules recommended in the statement allowed two major exceptions to the exclusion of pharmaceutical money from academic centers, even while conceding that industry might misuse the openings:

> Because the process of discovery and development of new drugs and devices often depends on input from academic medicine, consulting with or accepting research support from industry should not be prohibited. However, to ensure scientific integrity, far greater transparency and more open communication are necessary. . . . To promote scientific progress, AMCs should be able to accept grants for general support of research (no specific deliverable products) from pharmaceutical companies, provided that the grants are not designated for

> use by specific individuals. . . . As long as the institution stands
> between the individual investigator and the company making
> the grant, *the likelihood of undue influence is minimized but
> certainly not eliminated* [italics added].[9]

Recall that in response to the exposure of questionable business
dealings between NIH staff members and pharmaceutical firms, NIH
director Elias Zerhouni initially proposed an NIH-wide ban on *all* ties
to industry, including stock ownership in pharmaceutical firms by family members of NIH employees. In doing so, he responded to concerns
that in insidious ways industry's money can contaminate science. But
then he yielded to the reality that science is delivered to the public
by industry, as well as to warnings that the proposed rules rendered
the NIH less attractive in competition for outstanding scientists. Zerhouni eventually retreated from his draconian formula, confining the
restrictions to senior administrators, along with a relaxation of limits
on stock ownership by NIH staff members. Though unruly and troubled, and riddled with justified academic distrust of industry and an
accompanying wariness of academic purity regulations on the part of
industry, the coupling of industry and academic science is here to stay.

Old Sins Endure

The trend toward right behavior in that relationship, though favorable,
has been stimulated by failings that are far from eradicated, especially
in the clinical areas. Still plentiful, they include the old sins, going back
many years, encountered throughout this book: distortion or suppression of commercially inconvenient research findings, promiscuous patenting and the erosion of the scientific commons, clinical trials artfully
designed to produce favorable results for pharmaceutical manufacturers, concealment of financial dealings with industry by individual
researchers and academic institutions, and a variety of other transgressions. It is not yet time to demobilize the forces for scientific integrity.

Going beyond the narrowly focused federal regulations governing
scientific misconduct, revelations of offenses against scientific integrity
have propelled the academic research enterprise to adopt additional
safeguards, though with varying effects. Disclosure of financial conflicts of interest is now a common requirement for publication in mainstream scientific and medical journals and for university employment.
Disclosure does not ensure objectivity, and it may lull onlookers into
assuming that it neutralizes conflicts. But viewed with that understand-

ing, it nonetheless contributes to transparency, an indispensable, but not always sufficient, requirement for ensuring integrity in academic-corporate relations. At various other points in the research system, new safeguards have recently been installed. To counter the pharmaceutical industry's practice of concealing unfavorable clinical-trial findings, the most influential medical journals now require early registration of clinical trials in a public database as a prerequisite for consideration for publication. Clinical testing of drugs by faculty members with a financial stake in the outcome is prohibited by most universities, with minor exceptions permissible in special, carefully monitored circumstances. Through a rigorous accreditation and training program, serious efforts are in progress to raise the performance of the institutional review boards responsible for assuring the safety of volunteers in medical experiments. After a succession of appalling episodes, many if not most contracts between universities and corporate research sponsors guarantee scientists an unrestricted right to publish, with the exception of a grace period for the company to review the manuscript and retain proprietary information. Assignment of graduate students to their professors' commercial research is generally forbidden. Early in their training, most scientists now receive some formal instruction or mentoring in scientific ethics and responsible scientific behavior.

Anyone who doubts that progress has been made in the last four or five years is obstinately ignoring the beneficial changes that have occurred. But denial of reality also afflicts anyone who doubts that science has a long way to go in shedding questionable tactics and values, in protecting against backsliding, and in practicing transparency as an indispensable requirement for assuring scientific integrity. In support of the latter point, I'll relate a recent personal episode from the reportorial trail, involving the University of Pennsylvania, site of the death in 1999 of Jesse Gelsinger, age eighteen, in a botched clinical trial. If my experience, five years after that sad event, seems to conflict with my confidence that universities are educable in the cause of good behavior, I sadly concede that, yes, it does in this instance. Overall, however, I remain hopeful.

Penn's Got a Secret

While conducting interviews at the University of Pennsylvania in 2004, I heard of a recent agreement between Penn's School of Veterinary Medicine and Pfizer, Inc., the pharmaceutical firm, which produces veterinary products; also of another agreement, this one between Penn and IBM. My request to Penn officials for information about the university's agreements with the two companies brought the response that no announcements had been made at the outset concerning the deals and that no information would now be provided. I repeated my requests, in e-mails to several Penn officials, with specific questions about the university's arrangements with the two companies: "What are the terms of the agreements underlying these collaborations? What are the obligations of the parties? What are the financial arrangements?" In reply to my message, Leslie Hudson, at the time Penn vice provost and head of the university's Office of Strategic Initiatives, e-mailed me that "these are details which the university does not release without specific permission of our partners." A repeat request to Hudson brought the response, "I hope in the spirit of our helpfulness to date that you will characterize this as 'commercial terms that are not normally disclosed for these types of relationships.' " [10] A query to IBM produced an unfulfilled assurance that a response would be forthcoming. Pfizer, however, responded to my request for information.

In December 2005 a Pfizer spokesman, Robert Fauteux, group director for communications and public policy, telephoned me with details about his company's deal with Penn. Penn's reticence then became understandable. Under the terms of their collaboration, Fauteux told me, the Veterinary Clinical Investigation Center at Penn's School of Veterinary Medicine performs clinical trials on an experimental Pfizer drug for relieving cancer pain in dogs. For its part, he said, Pfizer pays half the salary of the director of the clinical center, half the salary of an administrative associate, and the full salary for a veterinary technician—all employees of the University of Pennsylvania.

Penn's guidance for pet owners interested in enrolling their pets in clinical trials at the veterinary school foggily addresses the issue of finance and conflict of interest: "To ensure no conflict of interest for the investigator conducting a trial," the guidance states, "our policy is that the investigator may not benefit financially from participating in a clinical trial from any source. Rather, the funding is used [to] pay for time, staff, and procedures to support and run the trial." [11] Pet owners were advised that additional information "about the funding of your trial" could be obtained from members of the staff.

Were Penn's recipients of Pfizer's salary money participating in the trial, and if so, wasn't that a financial benefit, I wondered. With the information from Pfizer in hand, I sought further information, asking Penn to confirm and explain the salary information and to advise me whether other salaries at Penn were paid by corporate sponsors of research. I was particularly interested in whether pharmaceutical firms paid any part of the salaries of Penn researchers conducting human clinical trials of their experimental drugs, and whether Penn researchers were allowed to conduct clinical trials of drugs in which they held a personal financial stake. A staff member in Penn's news office responded on January 10, 2006, that the salary information provided by Pfizer concerning the veterinary trials was correct and added, "As far as I am aware the arrangement is unique at Penn"—meaning, apparently, that it was confined to the veterinary school. No other information was provided in that response. The next day, however, a higher-level official at Penn, Vanda McMurtry, vice president for government affairs, asked me by e-mail to telephone him to discuss my inquiry. I repeated my questions about financing of clinical research. Explaining that he was relatively new on the job, McMurtry said he would seek an answer from other Penn officials. Weeks went by. After several prodding reminders from me, McMurtry replied, by e-mail, "I spoke with Perry [Molinoff, vice provost for research], and also with several other people here. I regret I don't have any information of the kind you are seeking."[12]

Penn, like scores of other universities, annually receives hundreds of millions of dollars of public money for research and other purposes. Almost unavoidably, any research activity on its campus benefits from or entails the use of publicly financed facilities. Nonetheless, the university—along with many others—conceals the terms of at least some of its deals with profit-seeking firms, even as it appeals for public trust and support. Is Penn embarrassed by the terms of its dealings with the two companies and perhaps other companies? The blackout of information suggests that it possibly is.

How extensive is such confidential dealing between universities and corporate partners? I don't know, but examples are not hard to find. In an interview at Stanford, Arthur Bienenstock, the university's vice provost and dean of research and graduate policy, declined to reveal the terms of a memorandum of understanding between Stanford and IBM for collaboration at a joint center for spintronics research. "Is the MOU [memo of understanding] made available to the university community?" I asked, reminding Bienenstock, an acquaintance from his service in the White House science office, that "this is the era of transparency, Artie."

"I understand that," Bienenstock replied. "But each side made concessions that it might not generally make because of each side's trust of the other to behave responsibly in difficult circumstances. We can't assume that in other relationships, and therefore we would be more protective. That's all I can say."

"More protective in what sense?" I asked.

"Indemnity, things like that," he replied, adding that "it's the absence of things in the contract that would be noted by some."

"So, it's a private agreement between the university and IBM, and that's all that's told to the world at large."

"Right," he replied.[13]

"An MD, Not an MDeity"

Sound rules for good behavior in science are in place. Though complex in some respects, they are not difficult for researchers to understand or at least raise concern and inspire a quest for guidance, if only to be safe. The great majority of scientists appear to play by the rules. But to an astonishing extent, and in seemingly brazen fashion, some don't. Years after the major scientific and medical journals mandated disclosure of financial connections that might entail conflicts of interest, they continued to encounter indifference and noncompliance among researchers submitting papers for publication. Few editors, however, were inclined to verify or even question the financial statements of their authors unless challenged—usually after publication—by other scientists. In 2006 *JAMA* acknowledged the omission of the requisite financial data in three prominent research reports that it had published that year, blaming the violation on the authors' failure to inform the editors of their prior business with drug manufacturers. Bearing the names of thirteen researchers, the first *JAMA* paper, published February 1, cautioned against discontinuing antidepressant treatment during pregnancy—a controversial conclusion of obvious cash interest to manufacturers of antidepressants. At the end of the article, two of the authors disclosed corporate connections. Their acknowledgment was followed by: "None of the other authors reported disclosures."[14] The *Wall Street Journal* subsequently reported that "at least seven of the others have [corporate] relationships that were not disclosed." The newspaper noted:

> The lead author—Lee S. Cohen, a Harvard Medical School professor and director of the perinatal and reproductive psychiatry research program at Massachusetts General Hospital—is a longtime consultant to three anti-depressant makers, a paid speaker for seven of them and has his research funded by four drug makers. None of his financial ties were reported

in the study. . . . Dr. Cohen and some of his coauthors subsequently hit the lecture circuit, telling physicians about their
findings while also spotlighting flaws in other recent studies
that have found increased risks to babies born to mothers who
use antidepressants.[15]

In this instance, and in the two other cases involving nondisclosure of financial connections, the authors contended that the research
reported in their papers was unrelated to their financial dealings with
pharmaceutical firms, and disclosure was therefore not required. Their
explanation challenged credulity, as Catherine D. DeAngelis, editor in
chief of *JAMA*, documented in an editorial titled "The Influence of
Money on Medical Science."[16] Starting in 1989, she wrote, *JAMA* required authors to disclose "the financial interests they have that might
be perceived as influencing the article they have written." In 1990
JAMA began publishing the disclosed information as appendages to
articles. In 1999, she noted, the disclosure requirement was expanded
to "any role the financial sponsor played in either the study or the
resulting article." Under an embarrassing spotlight, the nondisclosing
authors of the antidepressant report explained in a letter to *JAMA* that
financial data concerning their prior financial dealings with pharmaceutical firms was not provided because research for the disputed paper
was financed by the NIH. However, in a bow to the importance of
transparency, they expressed regret for the omissions and listed their
company connections—a long list, indeed, including pharmaceutical
firms that produce antidepressants.[17]

Criticism of *JAMA* in scientific circles and in the popular press also
elicited a remarkable expression of helplessness from editor DeAngelis.
In the same editorial, she argued, "There is simply no way to guarantee that all financial relationships and arrangements of all authors are
disclosed. It is not feasible to independently investigate the financial
relationships of every author, as no comprehensive, up-to-date source
of this information exists. Calling every author (for *JAMA*, that involves thousands of individuals annually) offers no advantage over our
current requirement that every author sign a document attesting to his
or her financial relationships or lack thereof. Misrepresentation of or
failure to completely disclose financial interests on the telephone or
in person is not much different than doing so in writing—in fact, one
might argue that requiring a signature better encourages honesty."

DeAngelis then candidly acknowledged a little-discussed reality of
scientific and medical publishing—the competition among journals for

groundbreaking, attention-getting papers that draw readers, who draw the advertisers that make the journals profitable for their owners, in most instances scientific and medical societies.

> Leveling sanctions against an author who fails to disclose financial interests by banning publication of his or her articles for some time period would only encourage that author to send his or her articles to another journal; it cleans our house by messing others. So what about all editors, or at least a group, such as the International Committee of Medical Journal Editors, agreeing to share the information and jointly to ban the offending authors? Those who suggest this approach have not considered the risk of an antitrust suit. Finally, the degree I hold is a MD, not an MDeity; I have no ability to know what is in the minds, hearts, or souls of authors. Furthermore, I do not have, nor desire to have, the resources of law enforcement agencies, but I do know that the accuracy of lie detector tests is questionable.

DeAngelis concluded by passing the responsibility for disclosure compliance by researchers to the institutions that employ them. "The most potent [tool]—both in enforcement and education—is the instigation of a full investigation by the deans of the authors' institutions. . . . In 2006, I have resorted to this approach twice, resulting in thorough investigations and appropriate corrective actions for the authors who were faculty members at the Mayo Clinic College of Medicine and the University of Nebraska School of Medicine, respectively."

At the end of the editorial, *JAMA*'s editor dutifully stated: "Financial Disclosures: None reported." [18]

Other editors have not passed the buck. Since 2004 *Environmental Health Perspectives* has adopted a three-year ban on violators of its disclosure requirements, the *Wall Street Journal* noted, while the *Journal of Thoracic and Cardiovascular Surgery* "said it would start to ban for 'some period of time' authors who fail to disclose conflicts." [19]

In Conflicted Circumstances, a Rare Resignation

Though disclosure of financial conflicts of interest has long been enshrined as a bedrock principle of ethical scientific behavior—especially when human subjects are involved—violations sometimes appear to be regarded as the scientific equivalent of jaywalking. In 2006 the jour-

nal *Neuropsychopharmacology,* published by the American College
of Neuropsychopharmacology, carried a favorable report on the use of
mild electric shock, delivered by an implanted device, for treatment of
depression.[20] The publication did not disclose that seven of the authors,
all academics, served as consultants to the company that manufactures
the device or that the eighth was an employee of the company. The lead
author of the article, Charles B. Nemeroff, chairman of the Department
of Psychiatry and Behavioral Sciences at Emory University, also served
as editor of *Neuropsychopharmacology.* Following the *Wall Street
Journal*'s disclosure of his conflicted role, he resigned the editorship.[21]

Such casual indifference to rules and regulations is not unusual in the
scientific enterprise. In 2001 newly issued federal regulations governing
the operations of institutional review boards stated that "no IRB may
have a member participate in the IRB's initial or continuing review
of any project in which the member has a conflicting interest, except
to provide information requested by the IRB."[22] In 2005 a nationwide
survey that brought responses from 574 IRB members found that
"15.1 percent reported that at least one protocol came before their IRB
during the previous year that was sponsored either by a company with
which they had a relationship or by a competitor of that company, both
of which could be considered conflicts of interest." Twenty-three percent of the members said they did not disclose the conflict.[23]

A Counter-Reformation?

Resistant to lay interference in its internal affairs, but protective of its
virtuous image, the scientific establishment has long been involved in
a two-step process of sins painfully exposed and reforms grudgingly
adopted though sometimes ignored. But something new has been
added: public attacks on the reforms. Professor Thomas Stossel published his screed against conflict-of-interest regulations (see page 171)
in the *New England Journal of Medicine,* long a leading site for deploring commercial excesses and ethical failures in research and medical
practice. His opinion was appropriately published as a contribution to
policy debate. But it should be recognized as an anti-regulatory straw
in the wind, an indication that a counter-reformation against regulatory strictures in research is gathering support.

Nature, also a fount of reports on scientific wrongdoing, recently published a news article headlined "Researchers Break the Rules in Frustration at Review Boards." The report summarized studies that concluded
that "some ethics panels are alienating researchers and inadvertently

promoting deceit" by insisting on strict adherence to regulations for the protection of experimental subjects. "Researchers acknowledge that the boards are necessary to ensure that subjects are treated correctly," the news report stated, "but sometimes complain that the boards fail to understand the research involved and do not explain their decisions properly." [24] In a recent survey, a considerable number of respondents acknowledged violations of basic rules of good scientific behavior. Even allowing for some ambiguity in the questions and answers, the survey results suggest considerable nonchalance toward good practice. Based on responses from 3,247 scientists, the survey found that 15.5 percent reported "changing the design, methodology or results of a study in response to pressure from a funding source." Ten percent acknowledged "inappropriately assigning authorship credit," while 0.3 percent admitted to both "falsifying or 'cooking' research data" and "ignoring major aspects of human-subject requirements." [25] Even where the reported percentages are small, the true numbers of individuals can be large because of the large size of the scientific enterprise.

Stubborn resistance, or outright indifference, to the regulation of research remains a durable aspect of the scientific enterprise, over a quarter of a century after the passage of the National Research Act and prior and succeeding mandates for good behavior in research. From inside and outside the scientific establishment, lamentations about failures to follow the rules increase in volume. But when it comes to asserting the imperatives of ethical scientific practice, the biomedical-research enterprise has frequently been revealed as impotent in its relations with industrial sponsors of clinical research. In 2002 a survey of clinical-research agreements between industry and 108 of the nation's 125 medical schools, published in the *New England Journal of Medicine,* concluded that

> academic institutions rarely ensure that their investigators have full participation in the design of the trials, unimpeded access to trial data, and the right to publish their findings. . . . The current research environment may impede institutions' attempts to negotiate contract provisions that secure investigators' rights. In response to some survey items, particularly those addressing publication and confidentiality, several respondents said they felt powerless in contract negotiations with sponsors. One respondent stated that although some institutions may be able to negotiate provisions that ensure investigators' rights, her institution was "just a small medical school." [26]

Three years later Dr. Steven E. Nissen, director of the Cleveland Clinic Cardiovascular Coordinating Center, looked back and concluded: "There has been little progress. The main reason is that commercial sponsors still exclusively control the database for most clinical trials."[27]

The extent of these failings is unknown, but they were plentiful enough as late as 2005 for the Association of American Medical Colleges to issue another homily urging scientists and their institutions to behave themselves. Fearful that academic science would be tarnished by complicity in Big Pharma's buccaneering methods of pushing drugs to market, the AAMC, in tandem with several other health-related organizations, coyly acknowledged that it "has been troubled by evidence that significant variation continues to exist within the academic community over the application of appropriate standards for analyzing and reporting the results of sponsored clinical research, especially clinical trials sponsored by industry." To make things right, the statement reiterated the fundamentals of sound clinical-research behavior, among them disclosure of financial conflicts of interest, prohibition of ghost authorship, registration of clinical trials, and prompt publication of full clinical-trial results.*[28]

The AAMC is rooted in the biomedical enterprise. In addressing the public, it usually tends toward optimism, insisting either that all is well in medical research, or, if amiss, will swiftly be corrected. That it felt the need to remind scientists of the rules of ethical behavior is revealing—and disturbing.

The same must be said of an impeccably wholesome declaration in behalf of scientific integrity issued in 2006 by the Federation of American Societies for Experimental Biology, which has expanded to twenty-two scientific societies with combined membership of over eighty-four thousand. FASEB, as we've seen, represents the working academic scientists, rather than their deans, presidents, or institutions. Two years in preparation by a distinguished committee, "Shared Responsibility, Individual Integrity: Scientists Addressing Conflicts of Interest in

---

*The continuing drumbeat of protests against Big Pharma's tactics may be having a beneficial effect on corporate candor. In December 2006, after reportedly spending nearly $1 billion in development costs, Pfizer abruptly abandoned a highly promising drug for heart disease, torcetrapib, when fatalities among those taking the drug in a clinical trial significantly exceeded those in a placebo control group. The drug was long touted as a blockbuster for replacing highly profitable Pfizer drugs nearing patent expiration. Unlike Merck and its efforts to conceal the dangers of Vioxx, Pfizer promptly announced the trial findings and dropped the drug. The next day Pfizer stock fell 10 percent.

Biomedical Research" reiterated the pieties of good scientific behavior that have been circulating through the halls of science for decades. Among nineteen "guiding principles" recommended for individual scientists:

> Investigators have a responsibility and commitment to conduct scientific activities objectively and with the highest professional standards.
>
> Investigators shall not enter into agreements with companies that prevent publication of research results.
>
> Investigators shall be aware of and adhere to individual journal policies on disclosure of industry relationships.
>
> Investigators shall not use federal funds to the benefit of a company, unless this is the explicit purpose of the mechanism used to fund the research (e.g., Small Business Innovation Research and similar grants).
>
> Investigators shall regard all significant financial interests in research involving human subjects as potentially problematic and thus requiring close scrutiny.

What is to be inferred about the ethical state of science from the declaration of these rudimentary principles in 2006?

The Dangers Ahead

Temptations for ethical lapses are abetted by institutional factors that are untamed. The academic arms race giddily accelerates. In Ponzi-scheme fashion, it inflames the pursuit of money for constructing research facilities needed to attract high-salaried scientific superstars who can win government grants to perform research that will bring glory and more money to the university. Academe's pernicious enthrallment by the rating system of *U.S. News & World Report* is a disgrace of modern higher education. But the pursuit of ratings and boasts of high standing in this peculiar sweepstakes persists. Steady state is unthinkable, an abandonment of the growth imperative that measures success and animates academic chieftains. Federal agencies provide only about 5 percent of the construction money for the annual ongoing additions of millions of square feet of laboratory space in universities, which means that other sources are strained to pay the costs.[29] An ugly secret of academic economics is that undergraduate tuition steadily rises above the rate of inflation to help finance the graduate science facilities and programs that bring luster to the big research universities. Little attention is given to the fact that the nationwide expansion of laboratory facilities for biotechnology, as well as other fields, is unmatched by increases in funds to support staff and research in the gleaming new buildings. The NIH, overwhelmingly the main support for the life sciences, is budget-becalmed by the Bush administration's fiscal wantonness, notwithstanding the president's promises of increased spending on research. The pharmaceutical industry has been cutting its own research budgets but even in affluent periods has never been a major source of money for university research. The exception was in clinical research, but, increasingly, drug trials are conducted abroad, where costs are lower and regulatory compliance is less easily monitored. Public universities have experienced a long decline in the proportion of their budgets provided by state governments. Is a financial famine about to descend on the scientific landscape? Possibly, given the strains in the public support of research. If it comes, scientific integrity will be threatened as never before by temptations to cut ethical corners to meet the needs of commercial patrons. In these circumstances, prudence calls for a moderation of the scientific-growth obsession that flourishes throughout academe. But that won't happen until financial scarcity leaves empty or unfinished laboratories on a few campuses, as a lesson in economy, politics, and reckless ambition for university trustees, managers, faculty, and boosters.

Don't Blame Bayh-Dole

Critics of science for sale trace major failings of scientific integrity to commercial pressure generated by the Bayh-Dole Act's requirement for universities to disclose and pursue the commercial potential of their government-supported research. But even if the act were repealed or substantially amended, neither of which appears likely, universities will continue to seek patents and commercial customers for their research, as many did prior to passage of the 1980 law. The difference now is that virtually all American universities of significant scale are either habituated to seeking commercial deals for their research or are trying to learn how. Around the world, with or without their own versions of Bayh-Dole, industrialized nations seek to emulate the academic-industrial linkages that have succeeded in the United States over the past twenty-five years. Though the rhetoric of technology transfer has undergone cosmetic alteration to deemphasize the indelicate pursuit of money in favor of social benefits, the potential for moneymaking remains a powerful force among university managers. The jackpots periodically struck by a small proportion of universities in the tech-transfer game nourish hopes that are unrealistic for the great majority of participants, but not impossible.

The costs of running even a modest tech-transfer office are substantial, and the returns are often negligible, though rationalized by the possibility of a bonanza down the road. The folly of maintaining many of these offices is apparent, but the odds are poor for persuading any university to outsource the function, as was done by some schools pre-Bayh-Dole through the services of the nonprofit Research Corporation. Some universities outsource review of their clinical trials to commercial institutional review boards. But the motivation there is to be rid of a bothersome function that can create internal antagonisms on campus while consuming the unpaid time of board members drawn from the faculty. The tech-transfer office, on the other hand, can be depicted as an attractive addition to the administrative superstructure. It's there to help faculty and the institution make money, even if most of them make very little.

Finding licensees for academic research is a difficult task that requires close collaboration between researchers and tech-transfer specialists knowledgeable about the state of science, the interests and needs of particular firms, and opportunities in the marketplace. Farming the tasks to an off-campus service juggling multiple academic clients is not as appealing as having an in-house shop, staffed with university employees and driven by the goal of enriching researchers and the institution. While some researchers shun the patent game, in defiance of Bayh-Dole's mandate to disclose, patent, and commercialize, their institutions are obediently in lockstep with the law's requirements, which are confined to federally financed research. However, in the realm of research collaborations that do not involve federal funds and therefore are not governed by Bayh-Dole, some innovative arrangements have sprouted in skeptical reaction to the enduring, often misguided emphasis on making money from academic patents.

Corrective Steps

In 2005, orchestrated by philanthropy's evangelist of entrepreneurship, the Ewing Marion Kauffman Foundation, a group of major research universities and high-tech firms, announced an "open source" collaboration in which "intellectual property . . . will be made available free of charge for commercial and academic use." Confined to information technology, in which the shelf life of products is often too short for the lengthy processes of patenting and negotiating licensing, the collaboration is to provide company money to the universities. Pradeep K. Khosla, the dean of engineering at one of the participants, Carnegie Mellon University, was quoted as saying, "I find this as very significant, as changing the mindset of universities and bringing us back to what we should be doing. Once you own patents, you start behaving like a company, and that's not what you should be doing." Noting that jackpots are rare in academic research, Khosla added that "and it's also not your mission to own patents, but it is your mission to educate students, to do great research and to take risks." In addition to Carnegie Mellon, the participants are big-league members of the academic-industrial complex: Georgia Institute of Technology; Rensselaer Polytechnic Institute; Stanford University; the University of California, Berkeley; the University of Illinois at Urbana-Champaign; and the University of Texas at Austin; the companies are Cisco, Hewlett-Packard, IBM, and Intel.[30]

The patent system, with its diminished criteria for patentability, is blamed for rewarding secretiveness, poaching on the scientific commons, loss of professional collegiality, and other manifestations of greed in science. The patent system is so out of whack that, fortunately, it's high on the agenda for congressional review, while several cases are pending in the U.S. Supreme Court. Sorely needed are stricter applications of the historical criteria for patenting: utility, novelty, and non-obviousness, all of which have been allowed to slide, to a large extent because of ill-considered budget cuts and a rising workload at the United States Patent and Trademark Office. The careless granting of patents has spawned long-running litigation over the use of research methods essential for the progress of science. Noting that "proprietary claims have increasingly moved upstream, from the end products themselves to the ground-breaking discoveries that made them possible in the first place," two legal scholars, Arti K. Rai and Rebecca S. Eisenberg, warn that "in the long run, the current system may, paradoxically, hinder rather than accelerate, biomedical research." The observation is valid, but the proposed remedy is probably politically unattainable—amendment of the sainted Bayh-Dole Act "to give [federal] funding agencies more latitude in guiding the patenting and licensing activities of their grantees."[31] But lacking other weapons against abuse of the patent system, the scientific community, through its journals and professional organizations, should not silently suffer grabs of intellectual property that impede research and run counter to the public interest.

They may not be able to declare these abuses illegal, but they can employ the shame weapon by publicly labeling them unclean. In the scientific culture, that hurts—and may possibly change behavior. Whether patent reform will be forthcoming soon, or at all, is not certain. And even less certain is the impact that various reform proposals would have on the ethical environment in academic research.

As Drummond Rennie, in our conversation, and many others have pointed out, lessening industry influence in clinical trials would be another important step in behalf of scientific integrity and public benefit. Several piecemeal measures have already been taken, such as mandatory registration of trials to prevent the disappearance of commercially unfavorable findings. Various proposals for further assuring untainted studies of the clinical effectiveness and economic worth of new drugs and devices have been floated in health-policy circles in recent years. At the institutional level, these include enhancement of the federal Agency for Healthcare Research and Quality, which has encountered stiff resistance from industry and parts of the medical profession; creation of a quasi-governmental clinical testing organization, akin to the major federally funded laboratories that universities and private-sector organizations manage for the Department of Energy and other government agencies; and contracting the task to the Institute of Medicine, the health-policy arm of the National Academy of Sciences.[32]

The pharmaceutical and biotech industries will deploy their formidable resources to thwart any effort to diminish their influence over clinical trials of the products they wish to market. Their political swat was plainly evident in the 2003 Medicare drug legislation, which explicitly prohibits the federal government from using its massive market power to negotiate drug prices. Calamities are effective for speeding regulatory reform—as the Thalidomide tragedy clearly illustrated. But reason, evidence, and foresight, rather than shocking body counts, should govern the rules of pharmaceutical benefits and safety. Professor Jerry Avorn, a Harvard Medical School professor who has written extensively for professional and popular audiences, asked in a recent article, "How can we capture . . . interest in less sensational problems of medication safety? A good start," he suggested, "would be to make a national commitment to publicly supported studies of drug risks so that no company could take possession of critical findings for its own purposes."[33] The goal is clear, the politics less so. But with mounting public antipathy toward the industry and the political realignment in Congress, the possibility has improved for an ethical upgrade in clinical testing.

Transparency

Rules and recommendations proliferate for right behavior in research. But for maintaining and strengthening scientific integrity, the most potent means is transparency, a much overused word that equates to visibility, openness, candor. To ensure right behavior in scientific affairs, transparency requires clear, firm rules, penalties for noncompliance, and knowledgeable observers. This should not be equated with spies in the lab. For a model, recall that the effective management of scientific misconduct successfully relies on easily understood criteria, established procedures for investigating reported offenses, and the good citizenship of scientists, technicians, and students in the scientific community. Management of moneymaking misdeeds in and around clinical research poses greater difficulties. But there is no justification for the pleas of unawareness or helplessness that are often sounded at important junctures when wrongdoing occurs in the scientific process. There are plenty of qualified observers at various places in the scientific enterprise who are capable of playing a role. Some have embraced the task of protecting the good name of science, while others are yet to report for duty. It was only after years of abuse that the editors of the most influential and prestigious medical and scientific journals agreed to crack down on undisclosed conflicts of interest, ghost authorship, repetitive publication to inflate the importance of research, and selective reporting of clinical trials. In the most vigorous of these efforts, authors are expected to state whether their papers are free of these afflictions. But compliance varies and enforcement is slack, though there's no dissent from the age-old faith that protection of the scientific literature from contamination is essential to the health, progress, and useful application of science. The customary peer reviews, by busy scientists serving as unpaid screeners, frequently amount to little more than a once-over quickly in search of conspicuous errors or unsubstantiated claims.

Editors protest that they lack the resources to do more to confirm the authenticity and accuracy of papers submitted for publication. Would occasional spot checks at research facilities by small visiting teams serve to elevate the care and integrity devoted to the preparation of research papers? The possibility of an on-the-scene review would probably have a broad, beneficial impact on the scientific enterprise. Perhaps the munificent NIH could shake loose a small sum to conduct a trial of this method for upgrading the scientific literature.

Without invoking the scrutiny of anti-trust authorities, individual journals, acting alone, could easily raise the level of disclosure compliance by publicly banishing an offender or two from their columns, temporarily or permanently. The Internet would instantly carry the news to the far corners of the scientific world. Competing journals might open their columns to the banished, but their editors would gain no glory from doing business with offenders against good scientific behavior. The scientific profession exalts reputation. Among scientists and journal editors, the risks of being classed as a rogue would have a wondrously beneficial effect on attention to the rules.

Illuminate the Marketplace

Business dealings between universities and industrial firms should be brought out of the dark through Internet posting of contracts and other agreements on publicly accessible data bases. Refusal to do so usually rests on the flimsy contention that legitimate proprietary information would be compromised. If that's the case, the proprietary information could be excluded from public disclosure, subject to confirmation by an independent review. Apart from legitimately secret matters, the details of commercial deals should be available to the campus community. Whether public or private, universities are *public* institutions, benefiting from tax-exempt, nonprofit status, large amounts of government research money, and many privileges because of the common assumption that they are dedicated to the public well-being. Private deals with a university unavoidably piggyback on the money and trust provided by the public. No rationale exists for concealing the details of these deals, except that it suits the preferences of business organizations and makes them more willing to provide money, which inclines universities to accept their terms. But that's no excuse.

Academe's involvement with commercial enterprise is bound to continue. The task, then, is to maximize the benefits of collaboration and minimize the risks and liabilities. Transparency is the first step toward this goal. Scientists and other faculty members should insist upon it, demanding open access to details of all commercial deals on campus, institutional as well as individual. Their professional societies should back them with codes of behavior and declarations of support when industrial sponsors seek to violate ethical norms. Trustees and university administrators should reconsider the pitfalls of unrestrained growth. And the press should do its part by sniffing out wrongdoing and activating the powers of the embarrassment factor. The existence of concealed consulting for many years at the NIH is a stain on the scientific and medical press. For bloggers and the young journalists who staff the campus and alternative newspapers that exist in and around our universities, academic-industrial dealings provide a wonderful, but generally neglected, opportunity for sharpening investigative skills—and keeping science and industry alert to the perils of behavior that can lead to embarrassing revelations.

The reactions I encountered from friends, colleagues, and acquaintances when they learned of the subject of this book indicate that academic science is viewed ambivalently. There's wide recognition that universities conduct pioneering research that leads to important tangible benefits for health, wealth creation, and defense. But also volunteered to me was the belief that unscrupulous, self-serving dealings are commonplace in academe's pursuit of money and renown from science. The first assessment is correct: Science is in good shape, productive and socially beneficial. The negative elements in science pose a more complicated, less measurable story. But as I've shown throughout this book, they are sufficiently prevalent to warrant concern for the good name and promise of science.

Epilogue
A Parable for Our Time

The swift ascent of the University of Avarice from obscurity to prominence in national academic standings was a seminal event of higher education in the second decade of the twenty-first century. The transformation was largely the work of an innovative leader in university affairs, Dr. Grant Swinger. As provost at one of America's leading universities, Swinger had long been prized by academic headhunters. Offered the presidency by the Avarice trustees, he promptly accepted and received clear marching orders to achieve national recognition for the little-known ninety-two-year-old institution. Details were not provided, but in the picture, too, according to several reports, was an extremely attractive compensation package.

Renowned in his previous academic position as the founder of the Center for the Absorption of Federal Funds, Swinger accepted the challenge and brought to it his characteristic boldness, acumen, and energy. The task was plainly daunting, leading one educational observer to liken it to "transforming a pig's ear into a silk purse." Nonetheless, few who knew Swinger doubted that he would mount a successful response to the challenge. Brisk in manner, confident, and decisive, he inspired trust through his leadership abilities and scholarly attainments. Early in his career, he was the precocious recipient of the prestigious Ripov Prize, awarded annually to the principal investigator with the largest number of concurrent research grants. Other major honors followed, firmly establishing Swinger as a high achiever in the linked realms of science and academic affairs.

Prior to Swinger's arrival, U Av, as it's known, had briefly attracted national attention for the body parts scandal at its School of Mortuary Science (SMS), which partnered with the business school in a prize-winning program to teach entrepreneurial skills to students and faculty. The episode was settled out of court in an agreement that provided unspecified payments to bereaved relatives in return for dropping the litigation, including challenges to several patents held by SMS. (By agreement of the parties, court records were sealed, but this did not prevent a supermarket tabloid from reporting the case in an article titled "On the Trail of Granny's Femur"). Otherwise, U Av existed in the nether zone of academe and science, along with numerous other postsecondary institutions that never register on the popular charts of university rankings. That changed for U Av with the arrival of President Grant Swinger.

Early Actions Win Praise

Barely settled into office, Swinger made national headlines by announcing a record-breaking fund-raising goal, an astonishing $10 billion. The figure was more than double the previous high mark, set in 2007 by Stanford University, a renowned magnet for donations in contrast to U Av's chronically paltry performance. At a press conference luncheon in the Rainbow Room atop Rockefeller Center, Swinger explained that the money would be "invested in excellence to meet the challenges that confront our nation and the world." Appointment of "distinguished faculty and concentration on urgent national problems" would have priority, Swinger said. He also disclosed that U Av would terminate the employment of adjunct teaching staff "because of the financial and professional insecurity of their positions." Replacements, he said, would come from the ranks of graduate students, thus providing "the next generation of academics with valuable teaching experience."

The *New York Times* reported Swinger's announcements in a front-page article headlined "Upstart U Reaches for the Stars—and the Big Bucks." In U Av's near-century-long existence, this was its only appearance in the *Times,* as well as in any other national publication, apart from matters related to the difficulties at the School of Mortuary Science, which was renamed the Institute for Physiology early in the Swinger administration. Inquiries about the progress of the fundraising drive were declined by U Av's rapidly expanding Office of Communications and Public Affairs as "premature, incompatible with privacy regulations, and potentially harmful to promising discussions with prospective donors."

U Av next came to broad public attention with another announcement, this one of record-breaking tuition and fees—$100,000 per academic year, including room and board, surpassing the prices of the great brand-name institutions of higher education. (Remember, we're now in the second decade of the twenty-first century, and the school charges reflect the accumulation of regular annual increases.) Applications for admission immediately soared, earning the heretofore obscure institution its first notice in the coveted ratings of *U.S. News & World Report.* "Hot newcomer," reported the acclaimed bible of academic rankings. *Newsweek* called it "Leapfrog U." The chairman of U Av's trustees indicated the board's satisfaction with its presidential choice in a congratulatory note to Swinger that simply stated: "We're on the way."

Swinger accompanied the tuition increase with another headline-winning announcement: Tuition would not only be reduced or eliminated for students from families with modest household incomes, but for the seriously impoverished, the university would actually provide payments to the students' families to compensate for earnings they might have provided if they went to work rather than to U Av. "We must be cognizant of today's economic realities," Swinger explained. *NBC News,* in its weekly feature "Making a Difference," hailed Swinger as "the bold leader of a new generation of academic statesmen, visionary in outlook, sensitive to individual and national needs, and determined to make a difference." NPR featured him in a searching on-air interview, for which he received many accolades. Inquiries about implementation of the generous tuition arrangements were dismissed as potential violations of privacy regulations.

Expanding the Horizons of Research

The second year of the Swinger presidency brought yet another innovation to U Av—the founding of a major research facility, the Hugo First Institute for Human Experimentation, designed, as Swinger explained at a groundbreaking ceremony, "to focus on 'translational research,'

EPILOGUE **289**

the formidable gap between basic science and bedside treatment." Noting that the National Institutes of Health had assigned a high priority and significant financial support for this type of research, Swinger vowed that the new institute would be "a leader in assuring the efficacy and safety of new treatments for the American people." Referring to institutional review boards (IRBs), the federally mandated panels that supervise the safety of research on humans at academic medical centers, Swinger observed that "as often as they protect patients, they also get in the way of medical progress and constructive relations between academic institutions and industrial organizations. At U Av," he said with a flourish, "IRB stands for something else: Here it means 'industrial research buddy.'" Buttons and T-shirts bearing those words were distributed to all laboratory staff members, and a similarly inscribed banner was hung above the entrance to the biochemistry building.

A statement praising Swinger was soon after issued by the Pharmaceutical Research and Manufacturers of America, comprised of the major pharmaceutical firms. The Biotechnology Industry Organization, which represents the smaller research-oriented firms on the frontiers of medical science, honored Swinger with its Someday Medal, awarded annually for "encouraging faith and investment in the therapeutic potential of genetic therapy." The *Wall Street Journal* editorialized that "the refreshingly straight-talking Dr. Swinger looks like a good prospect for taking the helm of the benighted Food and Drug Administration and steering it into the oblivion that it so richly deserves."

U Av's growing national prominence took a still-greater leap forward with the signing of an agreement with a major pharmaceutical firm that provided for the company's support of research in U Av's laboratory facilities, clinical testing of the firm's products at the Hugo First Institute, and patent sharing of promising developments. Financial and other details of the agreement were being withheld from public disclosure, U Av and the firm jointly announced, "in compliance with privacy regulations and the need to protect proprietary information in joint pursuit of therapeutic benefits for the American people." As was later revealed, in connection with these arrangements, Swinger became a paid consultant to the company and, along with several members of the institute, also received stock options in the firm. For strengthening the linkage between research and commercialization, the U Av's technology-transfer office was substantially expanded and its senior staff members were given tenured professorships. Swinger was honored for this innovation with a certificate of merit from the national association of technology transfer officials.

Initially, U Av's relationship with its pharmaceutical partner went smoothly, resulting in several promising patents and spin-offs financed by the company for commercial exploration and development of research that originated in U Av laboratories. But then the Swinger administration's seemingly unstoppable march to success encountered its first reverses.

Storm Clouds Emerge

From the teeming ranks of perpetually malcontent graduate students, postdoctoral fellows, idle former adjuncts, and other ingrates, several of the most disgruntled went public with a variety of grievances and allegations. Shielded by protections for so-called whistle-blowers, they risked little by doing so. The most damaging allegations contended that the Hugo First Institute routinely fabricated reports of clinical trials based on nonexistent experimental subjects; that in rare cases where trials were actually conducted, "results" were written up prior to administering the experimental drugs; and that research papers reporting the fraudulent trial results were routinely prepared by employees of the sponsoring firms and published under the names of U Av researchers. The complainants told tales of eradication of disappointing clinical trial data and drug sales pitches at continuing medical education programs by faculty members on the company payroll. Promptly pledging "full and complete transparency," President Swinger announced creation of "a blue-ribbon, independent inquiry." Following several closed-door meetings, the inquiry concluded that the allegations were "wholly without merit." Data corroborating the inquiry findings would not be made public, he announced, in conformity with privacy regulations and the need to protect proprietary information. To soothe feelings on campus, Swinger called for "an intergenerational dialogue concerning the new world of science," a move that won him further plaudits for leadership.

As the controversies over the Hugo First Institute and related matters seemed to recede, a new, serious difficulty unexpectedly arose. A student on a work-study assignment in the U Av development office during the staff's lunch break innocently answered a telephone inquiry from a reporter seeking information about progress toward the historic $10 billion fund-raising goal. The student, untutored in dealing with the press, helpfully explained, "They haven't gotten anything yet. Nothing. They're complaining all the time"—comments that were promptly published, with lightly veiled insinuations of setbacks in U

Av's progress toward national standing. Calling a press conference, Swinger earnestly pointed out that "major fund-raising is not an overnight process and is not amenable to penny-by-penny counting." At the same time, he announced that he would soon make an important announcement concerning a major development in the fund-raising campaign, thus inducing a wait-and-see caution in the press. Staff of the still-expanding Office of Communications and Public Affairs fanned out to warn reporters against premature conclusions. "Look out that you don't end up with egg on your face," or words to that effect, they advised the press. Attempts to contact the student who had spoken injudiciously of the fund-raising results brought the response that she was "no longer on campus," while several reporters were discreetly advised that further information might be obtained from the university's mental health clinic.

Though the Swinger administration was deeply troubled by these adversities, the complex and disputed details, local and difficult to verify, failed to travel far from campus. The glow of success around U Av remained undiminished, leading the various published rankings of academic quality to post even higher ratings for U Av. "Look Out, Ivies!" *U.S. News & World Report* declared.

Recruiters Come Calling

Grant Swinger had not achieved success and prominence by ignoring reality. U Av was taking on water, and he alone knew it. Thus, when a distinguished search committee approached him as a possible candidate for heading the newly established permanent National Commission on Scientific Integrity, Swinger didn't say no. Instead, citing the inviolability of his vow to shepherd U Av all the way to national greatness, he expressed appreciation for the proffered position, thereby whetting the committee's interest in him. When urged to consider the needs of the nation in comparison to those of a single institution, he modestly noted his relatively brief tenure as head of U Av, telling the aroused recruiters, "Perhaps at another time." Now ecstatic about the man in its sights, the search committee persisted and overcame his resistance. Expressing deep regret, Swinger informed the U Av trustees that "my sense of responsibility to our nation's scientific enterprise compels me to accept the challenge that has been thrust upon me."

In a statement accompanying the announcement of his appointment to head the commission, Swinger pledged "my deep commitment to nurturing and protecting scientific integrity. Nothing is more important to the well-being of our nation and the advancement of democracy." He promptly departed for Washington, where the *Washington Post* profiled him under the headline "New Ethics Cop on the Scientific Beat."

Meanwhile, U Av's jilted trustees sank into deep introspection. When all was going so well, they asked themselves, why would their prize president jump ship, leaving them leaderless and confronted by the many uncertainties inherent in finding a suitable successor? As a first step toward commencing the search, they ordered an audit of the books in expectation of finding a solid financial base that would help attract the next leader for continuing the climb to national prominence. Alas, the auditors' report was bleak: the $10 billion fund-raising campaign had not yet covered its expenses, let alone contributed to U Av's minuscule endowment; applications for admission had indeed risen in response to the record-breaking tuition level, but in the absence of funds for financial aid, enrollments were actually down. Further inquiry revealed that the Swinger regime had bequeathed other difficulties to U Av. Carrying their allegations to Washington, the aggrieved graduate students et al. received a warm reception on Capitol Hill, leading to an investigation of the Hugo First Institute by the Government Accountability Office. Pending the outcome of that inquiry, the NIH prudently froze all grants at the institute. At the same time, the firm collaborating with the institute, fearing for its reputation in the drug marketplace, invoked its contractual right to withdraw from the relationship, taking with it all intellectual property and several items of costly scientific apparatus. The general counsel for the firm reminded his counterpart at U Av that all dealings between the two organizations were protected by nondisclosure provisions.

Though busy organizing the newly created National Commission on Scientific Integrity, Grant Swinger maintained a careful watch on events at his prior place of employment and the related responses in Washington. Thus, even before his new office suite was completely furnished, he was not surprised by an urgent request to meet immediately with the chairman of the board of trustees of the commission. The meeting was brief and ended with the understanding that Dr. Swinger would have an opportunity to preview the press release announcing his resignation, but changes would be at the discretion of the board chairman.

Back at U Av, following a series of interviews, the trustees were particularly impressed by the strong professional credentials and demeanor of one candidate. In contrast to Grant Swinger's take-charge persona, she projected a calm self-assurance. Still recovering from the

Grant Swinger experience, the trustees were especially reassured when, asked for her philosophy of governance, she thoughtfully reflected for a moment and then replied, "Don't underestimate the power of greed in the halls of science or the wholesome presence of altruism and self-respect. And don't overlook shame and embarrassment as forces for good behavior in scientific affairs."

Without exception, the trustees reacted favorably to this sage formulation, though one of them of fleetingly thought he had previously encountered those words, perhaps in a recent book. But eager to get on with business, he joined with his colleagues in unanimously offering the presidency to this outstanding candidate, who promptly and graciously accepted.

Abbreviations

AAHRPP	Association for the Accreditation of Human Research Protection Programs
AAMC	Association of American Medical Colleges
AAU	Association of American Universities
ADP	adenovirus death protein
AEI	American Enterprise Institute
AHRQ	Agency for Healthcare Research and Quality
AMC	academic medical center
AUTM	Association of University Technology Managers
BIO	Biotechnology Industry Organization
BMS	Bristol-Myers Squibb
CRADA	cooperative research and development agreement
CRO	contract research organization
DCRI	Duke Clinical Research Institute
FASEB	Federation of American Societies for Experimental Biology
FDA	Food and Drug Administration
GAO	Government Accounting Office, renamed Government Accountability Office in 2004
HEW	Health, Education, and Welfare, U.S. Department of
HHS	Health and Human Services, U.S. Department of
IP	intellectual property
IRB	institutional review board
JAMA	*Journal of the American Medical Association*
NCI	National Cancer Institute
NIH	National Institutes of Health
NSF	National Science Foundation
OHRP	Office for Human Research Protections

OPRR   Office for Protection from Research Risks
ORI   Office of Research Integrity
PHRMA   Pharmaceutical Research and Manufacturers of America, the major
          research-oriented pharmaceutical firms, referred to as Big Pharma
PHS   Public Health Service
RO1   NIH grant for investigator-proposed research
SBIR   Small Business Innovation Research
SLU   Saint Louis University
STTR   Small Business Technology Transfer
UC   University of California
UCB   University of California, Berkeley
UCSF   University of California, San Francisco
VA   Veterans Administration
WARF   Wisconsin Alumni Research Foundation
WASH U   Washington University, St. Louis

Notes

INTRODUCTION

1. Daniel Bikle, interview with author, January 27, 2005.

2. Robert K. Merton, "Science and Democratic Social Structure," in *Social Theory and Social Structure*, enlarged edition (New York: Free Press, 1968), p. 613.

3. Christine Vogeli et al., "Data Withholding and the Next Generation of Scientists: Results of a National Survey," and David Blumenthal et al., "Data Withholding in Genetics and the Other Life Sciences: Prevalence and Predictors," *Academic Medicine* 81, no. 2 (February 2006): 128–36; 137–45.

4. Quoted in Stanton A. Glantz et al., *The Cigarette Papers* (Berkeley: University of California Press, 1996), p. 323.

5. "Academic Research and Development," in *Science and Engineering Indicators 2006* (Washington, D.C.: National Science Board, 2006), chap. 5.

6. Richard Carter, *Breakthrough: The Saga of Jonas Salk* (New York: Trident Press, 1966), p. 283.

CHAPTER ONE

1. Robert P. Kelch, interview with author, April 22, 2004.

2. Endowment date from "College Endowments Post 'Respectable' Returns for 2005," *Chronicle of Higher Education*, January 27, 2006; budget data from Web sites of cited universities.

3. John Hechinger, "When $26 Billion Isn't Enough," *Wall Street Journal*, December 17–18, 2005, data from National Association of College and University Business Officers.

4. *Chronicle of Higher Education,* "Careers," advertisement, September 2, 2005.

5. John L. Pulley, "Raising Arizona: Is Michael Crow's Remaking of a State University a Model, or a Mirage?" *Chronicle of Higher Education,* November 18, 2005.

6. "Top Fund Raisers," *Chronicle of Higher Education,* Almanac Issue, 2005–6, August 26, 2005.

7. *Chronicle of Higher Education,* advertisement, August 5, 2005.

8. Indirect-cost data provided to author by NIH Office of Communications and Public Liaison and NSF Office of Legislative and Public Affairs.

9. Warren Bennis and Hallam Movius, "Why Harvard Is So Hard to Lead," *Chronicle of Higher Education,* March 17, 2006.

10. *Survey of Research and Development Expenditures at Universities and Colleges,* fiscal year 2003 (Washington, D.C.: National Science Foundation, 2005), table A-6.

11. Princeton University, Office of Development, advertisement, *New York Times,* January 22, 2006.

12. Jocelyn Kaiser, "Panel Weighs Starter Ro1 Grants," *Science,* June 25, 2004, p. 1891.

13. Bruce Alberts, interview with author, September 16, 2004.

14. Kaiser, "Panel Weighs Starter Ro1 Grants," p. 1891.

15. Jeffrey Brainard, "NIH Program Seeks to Speed Grant Process for New Applicants," *Chronicle of Higher Education,* December 5, 2005.

16. "Earned Degrees Conferred, 2002–3," *Chronicle of Higher Education,* August 26, 2005.

17. "Q&A—White House Science Adviser John Marburger," *Science & Government Report,* July 15, 2006, p. 3.

18. Louis Uchitelle, "Columbia Gets Star Professor from Harvard," *New York Times,* April 5, 2002.

19. *Chronicle of Higher Education,* April 22, 2005, data from American Association of University Professors.

20. Scott Smallwood, "Faculty Salaries Up Only Slightly, Survey Finds," *Chronicle of Higher Education,* May 7, 2004.

21. Julie Shneyer, "Money, Prestige, and Titles: All in a Name. A Rose by Any Other Name Is Not as Lucrative for Columbia Researchers Seeking Grants," *Columbia Daily Spectator,* October 20, 2004.

22. Marvin G. Parnes, interview with author, April 22, 2004.

23. Joel Kirschbaum, interview with author, January 27, 2005.

24. *Chronicle of Higher Education,* November 22, 2002. Bok expanded on his thoughts in *Universities in the Marketplace: The Commercialization of Higher Education.* (Princeton, NJ: Princeton University Press, 2003).

25. "Median Salaries of Top College Administrators by Job Category and Type of Institution, 2004–5," *Chronicle of Higher Education,* August 25, 2005.

26. "Executive Compensation: What Leaders Make," *Chronicle of Higher Education,* November 18, 2005.

27. Karen W. Arenson and N. R. Kleinfield, "Columbia's President, an Expert on Free Speech, Gets an Earful," *New York Times,* May 25, 2005.

28. *Corporate Yellow Book,* Summer 2004 (New York: Leadership Directories, 2004).

29. *Chronicle of Higher Education,* Executive Compensation, November 24, 2006.

30. Karen W. Arenson, "Cornell President Resigns, Citing Split with Trustees," *New York Times,* June 12, 2005.

31. John Lippincott, quoted in "What Presidents Think about Higher Education, Their Jobs, and Their Lives," *Chronicle of Higher Education,* November 4, 2005.

32. *Chronicle of Higher Education,* Executive Compensation, November 24, 2006.

33. Michael Janofsky, "College Chief at American Agrees to Quit for Millions," *New York Times,* October 26, 2005.

34. U.S. Office of Research Integrity, *ORI Newsletter,* June 2004, p. 6.

35. *Chronicle of Higher Education,* May 27, 2005.

36. Piper Fogg, "When Private Colleges Come Knocking: As Those Institutions Get Richer, Public Universities Struggle to Keep Their Professors," *Chronicle of Higher Education,* November 12, 2004.

37. *New York Times,* May 8, 2005.

38. *Chronicle of Higher Education,* March 18, 2005.

39. *The Economist,* June 11, 2004.

40. "Recruitment," *Nature,* September 9, 2004, p. 1.

41. "Vision for the Future," Manchester University, http://www.manchester.ac.uk/aboutus/facts/vision/ (accessed January 2, 2006).

42. "Academic R&D Facilities and Equipment," in *Science and Engineering Indicators 2004* (Washington, D.C.: National Science Board, 2004), chap. 5.

43. National Science Foundation, "Universities Continue to Expand Their Research Space with the Largest Increase Since 1988," *InfoBrief* (NSF 05-314), Washington, D.C., June 2005.

44. Richard Smith, interview with author, April 23, 2005.

45. "Financing Universities: Mendicant Scholars," *Economist,* November 11, 2006.

46. John H. Barton, "Intellectual Property Rights: Reforming the System," *Science,* March 17, 2000, pp. 1933–34.

47. Eliot Marshall, "Depth Charges Aimed at Columbia's 'Submarine Patent,'" *Science,* July 25, 2003, p. 448.

48. Goldie Blumenstyk, "Columbia U. Nears End of Battle with Biotechnology Companies Over Patents," *Chronicle of Higher Education,* August 15, 2005.

49. Scot G. Hamilton, interview with author, November 11, 2003.

50. Goldie Blumenstyk, "Supreme Court Declines to Hear Rochester's Appeal in High-Stakes Patent Case," *Chronicle of Higher Education,* December 10, 2004.

51. Allyn Jackson, "Whatever Happened to Rochester? Two Years Later, Mathematics Is Getting Accolades," *Notices of the American Mathematical Society,* December 1997, pp. 1463–65.

52. Lou Berneman, interview with author, October 28, 2004.

53. Monsanto Company, "Monsanto Company, University of California Resolve Dispute Over Technology Used to Produce Bovine Somatotropin for Dairy Cows," press release, February 27, 2006.

54. Donald Kennedy, interview with author, February 19, 2004.

55. Harold Varmus, interview with author, December 16, 2004.

CHAPTER TWO

1. David Korn, interview with author, March 1, 2004.

2. National Science Foundation, *InfoBrief* (NSF 07-311), Washington, D.C., January 2007.

3. *AAAS Report XXX: Research and Development FY 2006* (Washington, D.C.: American Association for the Advancement of Science, 2005), pp. 32–33.

4. *Science and Engineering Indicators 2006* (Washington, D.C.: National Science Board, 2006), pp. 3–16.

5. *Academic Research and Development Expenditures: Fiscal Year 2004,* table 30, National Science Foundation (NSF 06-323), August 2006.

6. Jocelyn Kaiser, "Bristol-Myers Ends No-String Grants," *Science,* June 2, 2006, p. 1289.

7. "A Moment of Truth for America," advertisement, *Washington Post,* May 2, 1995.

8. *Science and Engineering Indicators 2006,* pp. 5–12.

9. "Top 100 Academic Institutions in R&D Expenditures, by Source of Funds: 2003," in *Science and Engineering Indicators 2006,* table 5-11.

10. Lita Nelsen, interview with author, September 24, 2004.

11. "Academic Research and Development," in *Science and Engineering Indicators 2006,* chap. 5.

12. Sheldon Krimsky, *Science in the Private Interest: Has the Lure of Profits Corrupted Biomedical Research?* (Lanham, MD: Rowman & Littlefield, 2003), p. 80.

13. Robert Lee Hotz, "Professors, Inc.," review of Krimsky, *American Scientist,* May–June 2004, pp. 282–83.

14. Jennifer Washburn, *University, Inc.: The Corporate Corruption of American Higher Education* (New York: Basic Books, 2005), p. 139.

15. Duke University, Office of News and Communications, memo to author, April 27, 2005.

16. "NIH Roadmap: Re-engineering the Clinical Research Enterprise," http://nihroadmap.nih.gov/clinicalresearchtheme/.

17. See n. 15.

18. "External Review of the Collaborative Research Agreement between Novartis Agricultural Discovery Institute, Inc. and the Regents of the

University of California," Institute for Food and Agricultural Standards, Michigan State University, July 13, 2004, p. 13.

19. American Association of University Professors, "Statement on Corporate Funding of Academic Research," November 2004, http://www.aaup .org/statements/Redbook/repcorf.htm.

20. Donald Kennedy, interview with author, February 19, 2004.

21. Goldie Blumenstyk, "Greening the World or 'Greenwashing' a Reputation?" *Chronicle of Higher Education*, January 10, 2003.

22. CBS News, "Energy Bigs Bankroll Stanford Research," *SciTech,* November 20, 2002.

23. Agreement for Global Climate and Energy Project (G-CEP), effective December 16, 2002, the Board of Trustees of the Leland Stanford Junior University, ExxonMobil Corporation, General Electric Company, Schlumberger Technology Corporation, Toyota Motor Corporation.

24. Susan Butts, external technology director, Dow Chemical Company, quoted in Virginia Gewin, "The Technology Trap: America's Widely-Admired System for Transferring Ideas from the Lab to the Marketplace Is Showing Signs of Distress," *Nature*, October 12, 2005, pp. 948–49.

25. Karl Koster, interview with author, September 24, 2004.

CHAPTER THREE

1. National Science Foundation, Division of Science Resources Studies, *National Patterns of R&D Resources* (Arlington, VA; biennial series, 2004).

2. Howard Bremer, "The First Two Decades of the Bayh-Dole Act as Public Policy," address to the National Association of State Universities and Land-Grant Colleges, Washington, D.C., November 11, 2001.

3. Ibid.

4. Quoted in Jon Sandelin, *30 Years of Innovation* (Northbrook, IL: Association of University Technology Managers, 2004).

5. "What Is a Patent?" Stanford University Office of Technology Licensing.

6. Michael G. Douglas, interview with author, April 27, 2005.

7. Texas A&M University system, "Regents Approve Patents and Commercialization of Research as New Consideration for Faculty Tenure," press release, May 26, 2006.

8. Sara Lipka, "Texas A&M Will Allow Consideration of Faculty Members' Patents in Tenure Process," *Chronicle of Higher Education*, May 30, 2006.

9. Carl Gulbrandsen, interview with author, April 27, 2004.

10. Karl Koster, interview with author, September 24, 2004.

11. William Brody, "From Minds to Minefields: Negotiating the Demilitarised Zone between Industry and Academia," remarks to the Biomedical Engineering Lecture Series, Johns Hopkins University, April 6, 1999.

12. Rhonda L. Rundle, "Johns Hopkins Backs Off Pact for Skin Care," *Wall Street Journal,* April 11, 2006.

13. Carl Gulbrandsen, interview with author, April 27, 2004.

14. *AUTM Licensing Survey: FY 2005, Survey Summary* (Northbrook, IL: Association of University Technology Managers, 2007).

15. Richard Jensen and Marie Thursby, "Proofs and Prototypes for Sale: The Licensing of University Inventions," *American Economic Review,* March 2001, pp. 250–59.

16. *Yale Daily News,* December 8, 2004.

17. *AUTM Licensing Survey: FY 2003* (Northbrook, IL: Association of University Technology Managers, 2004).

18. *AUTM, FY 2005.*

19. Sam Jaffe, "Ongoing Battle Over Transgenic Mice," *Scientist,* July 19, 2004.

20. Stephen Heuser, "Harvard Woos Firms to Fund Research," *Boston Globe,* November 9, 2005.

21. Clifton Leaf, "The Law of Unintended Consequences," *Fortune,* September 19, 2005, p. 250.

22. "Innovation's Golden Goose," *Economist,* December 14, 2002.

23. U.S. Patent and Trademark Office, *Patenting by Organizations,* 2004.

24. Andrew Pollack, "Market Place; Signs that Biotech Has Healthier Future," *New York Times,* April 4, 2006.

25. "Bayhing for Blood or Doling Out Cash? A Landmark Law Has Allowed American Universities to Profit by Patenting Their Innovations. But the Costs Are Adding Up," *Economist,* December 20, 2005.

26. Mary Sue Coleman, quoted in "Technology Managers Do Their Bit for World Health," letter to *Nature,* June 2, 2005, p. 561.

27. Birch Bayh, "A Time to Speak Up," preface to *The Better World Report: Technology Transfer Stories: 25 Innovations That Changed the World,* (Northbrook, IL: Association of University Technology Managers, 2006).

28. Katherine Ku, interview with author, January 31, 2005.

29. *Lambert Review of Business-University Collaboration: Final Report,* December 2003, report to HM Treasury; accessed at http://www.lambertreview.org.uk.

30. "Message from the President," in *AUTM Licensing Survey: FY 2004.*

31. Mark Crowell, in public discussion at AUTM 2004 meeting in Charleston.

32. George G. Harker III, interview with author, January 18, 2005.

33. John A. Fraser, interview with author, February 2, 2004.

34. Office of Technology Transfer, National Institutes of Health, Activities, FY 2003–2005; data also includes the Food and Drug Administration, a minor source of technology-transfer activity.

35. Goldie Blumenstyk, "Missed Chances: Even Successful Researchers Lose Out on the Big Bucks from Invention. Just Ask Jerome P. Horwitz," *Chronicle of Higher Education,* August 12, 2005.

36. Karen Erdman, "AIDS Drugs: Is the Government Research Program a Helping Hand for Patients, or a Handout for the Pharmaceutical Industry?" Public Citizen Health Research Group, *Health Letter,* May/June 1989, pp. 10–17.

37. Public Citizen Health Research Group, *Health Letter,* Washington, D.C., August 2003.

38. Steven M. Ferguson, interview with author, December 1, 2004.

39. Chuck Ludlam, interview with author, October 22, 2004.

40. Harold Varmus, interview with author, December 16, 2004.

41. President's Council of Advisors on Science and Technology, *Report on Technology Transfer of Federally Funded R&D: Findings and Proposed Actions,* May 12, 2003.

42. Lita Nelsen, interview with author, September 24, 2004.

43. David C. Mowery, Richard R. Nelson, Bhaven N. Sampat, and Arvids A. Ziedonis, *Ivory Tower and Industrial Innovation: University-Industry Technology Transfer Before and After the Bayh-Dole Act* (Stanford: Stanford University Press, 2004), p. 179.

44. Mowery et al., *Ivory Tower,* pp. 184–85.

45. Royal Society, *Keeping Science Open: The Effects of Intellectual Property Policy on the Conduct of Science,* Policy Document 02/03, April 2003, London.

46. Donald Kennedy, editorial, *Science,* August 23, 2002, p. 1237.

47. Mowery et al., *Ivory Tower,* p. 183.

48. Christophe Lécuyer, "What Do Universities Really Owe Industry?: The Case of Solid State Electronics at Stanford," *Minerva* 43, no. 1 (2005): 51–71.

49. Daniel Lee Kleinman and Steven P. Vallas, "Contradiction in Convergence: Universities and Industry in the Biotechnology Field," in *The New Political Sociology of Science: Institutions, Networks, and Power,* edited by Scott Frickel and Kelly Moore (Madison: University of Wisconsin Press, 2006).

CHAPTER FOUR

1. Richard Burgess, interview with author, April 28, 2004.

2. Lisa Jarvis, "P&G to Drop Pharma R&D: Strategy Shift to In-Licensing Will Cost Scores of Science-Related Jobs," *Chemical & Engineering News,* February 27, 2006, p. 6.

3. Richard Bruno, interview with author, October 5, 2004.

4. Bruce Alberts, interview with author, September 16, 2004.

5. Harold Varmus, interview with author, December 16, 2004.

6. Ewing Marion Kauffman Foundation, "Kauffman Foundation Challenges Universities to Institutionalize Entrepreneurship on Campus," press release, July 7, 2003.

7. Institute of Education Sciences, *Digest of Education Statistics Tables and Figures,* 2005 Tables, National Center for Education Statistics, U.S. Department of Education, 2006.

8. "2004 Survey of Endowed Positions in Entrepreneurship and Related Fields in the United States," by J. A. Katz and Associates, for the Ewing Marion Kauffman Foundation, 2004.

9. Ewing Marion Kauffman Foundation, "Kauffman Campuses Collegiate Entrepreneurship," press release, 2005.

10. University of Rochester News, "Rochester Graduates Hear Words of Advice for Today," press release, May 15, 2005.

11. *Chronicle of Higher Education, Almanac Issue 2005–6,* August 26, 2005; and *Academic Research and Development Expenditures: Fiscal Year 2004,* Table 30, National Science Foundation (NSF 06-323), August 2006.

12. *AUTM Licensing Survey: FY 2003* (Northbrook, IL: Association of University Technology Managers, 2004).

13. Ewing Marion Kauffman Foundation, "Washington University in St. Louis: Executive Summary—Kauffman Campuses Proposal," press release, July 2, 2005.

14. Kauffman Fellowship, announcement, Division of Biology and Biomedical Sciences, Washington University School of Medicine.

15. "Accomplishments: Skandalaris Center for Entrepreneurial Studies," Washington University, February 2005. Report prepared for the Entrepreneurship Council.

16. Ken Harrington, interview with author, April 28, 2005.

17. Michael G. Douglas, interview with author, April 27, 2005.

18. *Chronicle of Higher Education,* advertisement, September 2, 2005.

19. Marvin G. Parnes, interview with author, April 22, 2004.

20. Robert Kelch, interview with author, April 22, 2004.

21. Charles L. Liotta, interview with author, January 19, 2005.

22. Roger P. Webb, Interview with author, January 19, 2005.

23. Steven Danyluk, interview with author, January 19, 2005.

24. Joel Kirschbaum, e-mail to author, February 28, 2005.

25. David Korn, interview with author, March 1, 2004.

CHAPTER FIVE

1. David Goldston, remarks, symposium on "The Future of the Triple Helix: Finding the Balance among Government, Industry, and Academic Research Relationships," AEI-Brookings Joint Center for Regulatory Studies, Washington, D.C., June 10, 2004.

2. Dennis E. Baker, FDA associate commissioner for regulatory affairs, letter to James M. Wilson, University of Pennsylvania, February 8, 2002; online at http://www.fda.gov/foi/nooh/Wilson.htm.

3. Scott Hensley, "Targeted Genetics' Genova Deal Leads to Windfall for Researcher," *Wall Street Journal,* August 9, 2000.

4. Rick Weiss, "U.S., Researchers Reach Deal in '99 Gene Therapy Case," *Washington Post,* February 10, 2005.

5. David Korn, interview with author, March 1, 2004.

6. *Guidelines for Dealing with Faculty Conflicts of Commitment and Conflicts of Interest in Research,* (Washington, D.C.: Association of American Medical Colleges, 1990).

7. Quoted in David Willman, "Stealth Merger: Drug Companies and Government Medical Research," *Los Angeles Times,* December 7, 2003.

8. *Outside Activities of Senior-Level NIH Employees* (OEI-01-04-00150), Office of Inspector General, Department of Health and Human Services, July 2005.

9. *Report of the National Institutes Blue Ribbon Panel on Conflict of Interest Policies*, A Working Group of the Advisory Committee to the Director, National Institutes of Health; draft: May 5, 2004.

10. Lobbying and campaign-contribution data from Center for Responsive Politics, http://opensecrets.org.

11. William M. Welch, "Tauzin Switches Sides from Drug Industry Overseer to Lobbyist," *USA Today*, December 15, 2004.

12. Jocelyn Kaiser, "NIH Chief Clamps Down on Consulting and Stock Ownership," *Science*, February 11, 2005, pp. 824–25.

13. NIH director Elias Zerhouni, letter to Chairman Joe Barton, Committee on Energy and Commerce, U.S. House of Representatives, July 8, 2005.

14. Bruce Alberts, interview with author, September 16, 2004.

15. David Willman, "The National Institutes of Health: Public Servant or Private Marketer?" *Los Angeles Times*, December 22, 2004.

16. Quoted in ibid.

17. Stephen Katz, testimony, hearing on "Avoiding Conflicts of Interest at the National Institutes of Health," U.S. Senate Committee on Appropriations, Subcommittee on Labor, Health and Human Services, Education and Related Agencies, January 22, 2004.

18. *Report of the National Institutes of Health Blue Ribbon Panel on Conflict of Interest Policies.*

19. Jocelyn Kaiser, "Elias Zerhouni: Taking Stock," *Science*, June 3, 2005, pp. 1398–99.

20. *Report of the National Institutes Blue Ribbon Panel on Conflict of Interest Policies.*

21. "Avoiding Conflicts of Interest at the NIH," statement of Ruth Kirschstein, senior advisor to the director, National Institutes of Health, submitted to U.S. Senate Committee on Appropriations, Subcommittee on Labor, Health and Human Services, Education, and Related Agencies, January 22, 2004.

22. "The Policies and Procedures in Place to Avoid Conflicts of Interest at the National Institutes of Health," statement of Marilyn L. Glynn, acting director, Office of Government Ethics, submitted to U.S. Senate Committee on Appropriations, Subcommittee on Labor, Health and Human Services, Education, and Related Agencies, January 22, 2004.

23. *Report of the National Institutes of Health Blue Ribbon Panel on Conflict of Interest Policy.*

24. Gardiner Harris, "A Businessman-Scientist Shakes Up the Health Institutes," *New York Times*, August 30, 2005.

25. *Report of the National Institutes of Health Blue Ribbon Panel on Conflict of Interest Policy.*

26. "Too Strict at NIH," editorial, *Washington Post*, February 23, 2005.

27. "Taking a Hard Line on Conflicts," editorial, *Nature*, February 10, 2005, p. 557.

28. Association of American Medical Colleges, IRS 990, "Return of Organization Exempt from Income Tax," 2002, accessed from http://www.Guidestar.com.

29. Jordan J. Cohen, "Restoring Public Trust in the NIH: Lessons for Academia," *AAMC Reporter,* March 2005, p. 2.

30. Memo circulated to NIH staff, "Conflict of Interest Information and Resources: Questions and Answers for Employees," National Institutes of Health, August 25, 2005.

31. David Willman, "Federal Scientist Admits He's Guilty of Conflict of Interest," *Los Angeles Times,* December 9, 2006.

32. Rick Weiss, "NIH Punishments Criticized: Members of Congress Complain that Sanctions Are Too Soft," *Washington Post,* September 14, 2006.

33. Kaiser, "Elias Zerhouni: Taking Stock," pp. 1398–99.

CHAPTER SIX

1. Greg Koski, interview with author, May 6, 2005.

2. CFR, Title 45, Part 46, Subpart A, Sections 46.111 and 46.116, "General Requirements for Informed Consent."

3. Task Force on Financial Conflicts of Interest in Clinical Research, *Protecting Subjects, Preserving Trust, Promoting Progress: Policy Guidelines for the Oversight of Individual Financial Interests in Human Subjects Research,* (Washington, D.C.: Association of American Medical Colleges, December 2001).

4. David A. Lepay, director, Division of Scientific Investigations, FDA, quoting an unnamed source, Conference on Human Subject Protection and Financial Conflicts of Interest, NIH, August 15–16, 2000.

5. NIH Grants Policy Statement, Part II, Terms and Conditions of NIH Grants Awards, March 2001.

6. Jeremy Sugarman et al., "The Cost of Institutional Review Boards in Academic Medical Centers," *New England Journal of Medicine,* April 28, 2005, pp. 1825–27.

7. Jeremy Sugarman, remarks to Conference on Human Subject Protection and Financial Conflicts of Interest, NIH, August 15–16, 2000.

8. Gary Ellis, conversation with author, May 19, 1999, quoted in Daniel S. Greenberg, *Science, Money, and Politics* (Chicago: University of Chicago Press, 2001), p. 360.

9. Lindsay A. Hampson et al., "Patients' Views on Financial Conflicts of Interest in Cancer Research Trials," *New England Journal of Medicine,* November 30, 2006, pp. 2330–37.

10. Sharon K. Friend, interview with author, January 26, 2005.

11. Quoted in *Implementing Human Research Regulations: The Adequacy and Uniformity of Federal Rules and of Their Implementation,* President's Commission for the Study of Ethical Problems in Medicine and Biomedical Behavioral Research, March 1983, p. 12.

12. *Scientific Research: Continued Vigilance Critical in Protecting Human Subjects* (GAO/HEHS-96-72, 1996).

13. Jordan Cohen, "IRBs: 'In Jeopardy,' or in Need of Support?" *AAMC Reporter,* August 1998, p. 2.

14. Thomas Bodenheimer, "Uneasy Alliance—Clinical Investigators and the Pharmaceutical Industry," *New England Journal of Medicine,* May 18, 2000, pp. 1539–44.

15. Mark R. Yessian, regional inspector general for evaluations and inspections, U.S. General Accounting Office, testimony to Committee on Government Reform, Subcommittee on Criminal Justice, Drug Policy and Human Resources, U.S. House of Representatives, December 9, 1999.

16. Ibid.

17. Thomson CenterWatch, memo to author, October 2005, derived from various Thomson reports.

18. "Harvard Keeps Strict Rules on Outside Research Work," *New York Times,* May 25, 2000.

19. *Wall Street Journal,* June 24, 1999.

20. "MIT Policies and Procedures: Faculty Rights and Responsibilities: Conflict of Interest," online at http://web.mit.edu/policies/4.4.html.

21. Derek Bok, *Universities in the Marketplace: The Commercialization of Higher Education* (Princeton, NJ: Princeton University Press, 2003), p. 21.

22. Perry Molinoff, interview with author, October 28, 2004.

23. Regis Kelly, interview with author, January 25, 2005.

CHAPTER SEVEN

1. Donna Shalala, remarks at conference on "Renegotiating the Grand Bargain: People, Health and the Pharmaceutical Industry—Who Owes What to Whom," November 3, 2005, sponsored by the Academy of Pharmaceutical Physicians and Investigators Education Foundation, Philadelphia.

2. J. Thomas Puglisi, director, Division of Human Subject Protections, Office for Protection from Research Risks, testimony, U.S. House Committee on Veterans Affairs, Subcommittee on Oversight and Investigations and Subcommittee on Health, April 21, 1999.

3. Jennifer Proctor, "Responding to Restrictions on Research," *Reporter,* Association of American Medical Colleges, April 2000, p. 1.

4. Bruce Alberts, interview with author, September 16, 2004.

5. Rick Weiss, "U.S. Halts Human Research at Duke," *Washington Post,* May 12, 1999.

6. Quoted in Proctor, "Responding to Restrictions on Research," p. 1.

7. Center for Information and Study on Clinical Research Participation, "101 Facts about Clinical Research," http://www.ciscrp.org/information/facts.asp (accessed February 23, 2006).

8. R. D. Start et al., "Evaluating the Reliability of Causes of Death in Published Clinical Research," *British Medical Journal,* January 25, 1997, p. 271.

9. Quoted in ibid.

10. National Academies, "Preventing Death and Injuries from Medical Errors Requires Dramatic, System-Wide Changes," press release, November 29, 1999.

11. *Science & Engineering Indicators—2002* (Washington, D.C.: National Science Board, 2002), vol. 1, pp. 7–12.

12. *Federal R&D Funding by Budget Function, Fiscal Years 2001–03, Special Report,* Division of Science Resource Statistics, National Science Foundation, August 2002.

13. Steven A. Goldstein, interview with author, April 23, 2004.

14. "Report to the Advisory Committee to the Director, NIH, from the Office for Protection from Research Risks Review Panel," June 3, 1999.

15. Donna Shalala, "Protecting Research Subjects—What Must Be Done," *New England Journal of Medicine,* June 14, 2000, pp. 808–10.

16. Greg Koski, remarks at conference on "Human Subject Protection and Financial Conflicts of Interest," NIH, August 15–16, 2000.

17. Jordan Cohen, "Trust Us to Make a Difference," address to the annual meeting of the Association of American Medical Colleges, October 29, 2000.

18. *Biomedical Research: HHS Direction Needed to Address Financial Conflicts of Interest,* U.S. General Accounting Office (GAO-02-89), November 2001.

19. David Armstrong, "Cleveland Clinic Had Ties to Maker of Faulted Device," *Wall Street Journal,* December 16, 2005.

20. "Draft Interim Guidance: Financial Relationships in Clinical Research: Issues for Institutions, Clinical Investigators, and IRBs to Consider When Dealing with Issues of Financial Interests and Human Subject Protection," http://ccnmtl.edu/projects/rcr/rcr_conflicts/misc/Ref/OHRP_CoI.pdf (accessed December 2, 2005).

21. FASEB letter to the Department of Health and Human Services, September 29, 2000, http://opa.faseb.org/pdf/9x29xooltr.pdf (accessed November 14, 2006).

22. *Report on University Protections of Human Beings Who Are the Subjects of Research,* Report and Recommendations, Task Force on Research Accountability, Association of American Universities, Washington, D.C., June 28, 2000.

23. Quoted in Eliot Marshall, "Universities Puncture Modest Regulatory Trial Balloon," *Science,* March 16, 2001, p. 2060.

24. Association of American Medical Colleges, "Supporters of Clinical Research Reaffirm Commitment to the Protection of Patients," press release, June 8, 2000.

25. Johns Hopkins University, "Report of Internal Investigation into the Death of a Volunteer Research Subject," July 2001, http://www .hopkinsmedicine.org/press/2001/july/report_of_internal_investigation.htm (accessed September 2005).

26. Greg Koski, interview with author, May 6, 2005.

27. *NewsHour with Jim Lehrer,* July 20, 2001.

28. Ibid.

29. Ibid.

30. Greg Koski, testimony, Committee on Veterans' Affairs, Subcommittee on Oversight and Investigations, U.S. House of Representatives, June 18, 2003.

31. *Protecting Subjects, Preserving Trust, Promoting Progress: Policy and Guidelines for the Oversight of Individual Financial Interests in Human Subjects Research,* Task Force on Financial Conflicts of Interest in Clinical Research, Association of American Medical Colleges, Washington, D.C., December 2001.

32. Ibid.

33. *Report on Individual and Institutional Financial Conflict of Interest,* Report and Recommendations, Task Force on Research Accountability, Association of American Universities, Washington, D.C., October 2001.

34. *Protecting Subjects, Preserving Trust, Promoting Progress II: Principles and Recommendations for Oversight of an Institution's Financial Interests in Human Subjects Research,* Task Force on Financial Conflicts of Interest in Clinical Research, Association of American Medical Colleges, October 2002.

35. Tape-recorded and transcribed by author.

36. Thomas P. Stossel, "Regulating Academic-Industrial Research Relationships—Solving Problems or Stifling Progress?" *New England Journal of Medicine,* September 8, 2005, pp. 1060–65.

37. Susan Ehringhaus and David Korn, *U.S. Medical School Policies on Individual Financial Conflicts of Interest: Results of an AAMC Survey,* Association of American Medical Colleges, 2004, http://www.aamc.org/research/coi/start.htm (accessed December 2005).

38. Greg Koski, interview with author, May 6, 2005.

39. AAHRPP, "Three Universities Become the First in Their States to Win Accreditation for the Way They Protect Research Participants," press release, September 21, 2005.

40. Bernard Schwetz, interview with author, December 17, 2003.

41. "Financial Relationships and Interests in Research Involving Human Subjects: Guidance for Human Subject Protection," *Federal Register,* May 12, 2004, pp. 26393–97.

CHAPTER EIGHT

1. For summarizing the long and controversy-ridden tale of the origins and marketing of Taxol, I have relied heavily on two excellent sources: Frank Stephenson, "A Tale of Taxol," Florida State University, *Research in Review* (Fall 2002), pp. 12–37; and *Technology Transfer: NIH-Private Sector Partnership in the Development of Taxol,* United States General Accounting Office (GAO-03-829), June 2003.

CHAPTER TEN

1. Kathleen M. Farrell, interview with author, April 29, 2005.

2. Jennifer Washburn, *University, Inc.: The Corporate Corruption of American Higher Education* (New York: Basic Books, 2005), p. 97.

3. Kathleen Farrell, e-mail to author, January 10, 2006.

CHAPTER ELEVEN

1. Eliot Marshall, "Startling Revelations in UC-Genentech Battle," *Science,* May 7, 1999, pp. 883–86.

2. UCSF Facts and Figures, http://www.ucsf.edu/about_ucsf/profile.html.

CHAPTER THIRTEEN

1. Kay Dickersin and Drummond Rennie, "Registering Clinical Trials," *JAMA,* July 23/30, 2003, pp. 516–23.

2. The ALLHAT Study, on Antihypertensive and Lipid-Lowering Treatment to Prevent Heart Attack. Randall S. Stafford et al., "Impact of Clinical Trial Results on National Trends in %-Blocker Prescribing, 1996–2002," *JAMA,* January 7, 2004, pp. 54–62.

CHAPTER FOURTEEN

1. Daniel E. Koshland, editorial, "Fraud in Science," *Science,* January 9, 1987, p. 141.

2. Office of Research Integrity, "About ORI—History," http://ori.dhhs.gov/about/history.shtml (accessed March 25, 2006).

3. Appendix Table 5-23, "U.S. and World Scientific and Technical Articles, by Field: 1973–86," in *Science & Engineering Indicators—1989* (Washington, D.C.: National Science Board, 1989).

4. Adam Silverman, "Former UVM Professor Sentenced to Jail for Fraud," *Burlington Free Press,* June 29, 2006. The case is examined at length by Jeneen Interlandi, "An Unwelcome Discovery," *New York Times Magazine,* October 22, 2006.

5. Data by e-mail from Lawrence J. Rhoades, director, Education and Integrity Division, Office of Research Integrity, March 30, 2006.

6. University of Pittsburgh, "Summary Investigative Report on Allegations of Possible Scientific Misconduct on the Part of Gerald P. Schatten, Ph.D.," February 8, 2006.

7. Steven Shapin, *A Social History of Truth: Civility and Science in Seventeenth-Century England* (Chicago: University of Chicago Press, 1994), p. 413.

8. "Pharma Must Earn—Not Buy—Its Reputation," *Medical Marketing & Media,* January 2006, p. 30.

9. Troyen A. Brennan et al., "Health Industry Practices That Create Conflicts of Interest: A Policy Proposal for Academic Medical Centers," *JAMA,* January 25, 2006, pp. 429–33.

10. Leslie Hudson, e-mail to author, December 9, 2004; January 6, 2005.

11. University of Pennsylvania, School of Veterinary Medicine, Department of Clinical Studies, Philadelphia, Research, http://www.vet.upenn.edu/departments/csp/research/clinical_trials/ (accessed January 11, 2006).

12. Vanda McMurtry, e-mail to author, February 9, 2006.

13. Arthur Bienenstock, interview with author, January 28, 2005.

14. Lee S. Cohen et al., "Relapse of Major Depression During Pregnancy in Women Who Maintain or Discontinue Antidepressant Treatment," *JAMA,* February 1, 2006, pp. 499–507.

15. David Armstrong, "Financial Ties to Industry Cloud Major Depression Study," *Wall Street Journal,* July 11, 2006.

16. Catherine D. DeAngelis, "The Influence of Money on Medical Science," *JAMA,* August 7, 2006, pp. 296–98.

17. Lee S. Cohen et al., letter to the editor, *JAMA,* July 12, 2006, p. 296.

18. DeAngelis, "The Influence of Money," pp. 296–98.

19. David Armstrong, "Medical Reviews Face Criticism Over Lapses," *Wall Street Journal,* July 19, 2006.

20. Charles B. Nemeroff et al., "VNS Therapy in Treatment-Resistant Depression: Clinical Evidence and Putative Neurobiological Mechanisms," *Neuropsychopharmacology,* July 2006, pp. 1345–55.

21. David Armstrong, "Medical Journal Editor Nemeroff Steps Down Over Undisclosed Ties," *Wall Street Journal,* August 28, 2006.

22. # 45 CFR, 46.107 (e), 2001.

23. Eric G. Campbell et al., "Financial Relationships between Institutional Review Board Members and Industry," *New England Journal of Medicine,* November 30, 2006, pp. 2321–29.

24. Jim Giles, "Researchers Break the Rules in Frustration at Review Boards," *Nature,* November 10, 2005, pp. 136–37.

25. Brian C. Martinson et al., "Scientists Behaving Badly," *Nature,* June 9, 2005, pp. 737–38.

26. Kevin A. Schulman et al., "A National Survey of Provisions in Clinical Trial Agreements Between Medical Schools and Industry Sponsors," *New England Journal of Medicine,* October 24, 2002, pp. 1335–41.

27. Quoted in Robert Steinbrook, "Gag Clause in Clinical-Trial Agreements," *New England Journal of Medicine,* May 26, 2005, pp. 2160–62.

28. Association of American Medical Colleges, "Principles for Protecting Integrity in the Conduct and Reporting of Clinical Trials," September 15, 2005.

29. *Science and Engineering Indicators 2006* (Washington, D.C.: National Science Board, 2006).

30. Scott Jaschik, "New Model for Tech Transfer," *Inside Higher Ed,* December 22, 2005. http://insidehighered.com/news/2005/12/22/open.

31. Arti K. Rai and Rebecca S. Eisenberg, "Bayh-Dole Reform and the Progress of Biomedicine," *American Scientist,* January–February 2003, pp. 52–59.

32. These and other proposals are discussed in Gail R. Wilensky, "Developing a Center for Comparative Effectiveness Information," *Health Affairs,* November 7, 2006.

33. Jerry Avorn, "Dangerous Deception—Hiding the Evidence of Adverse Drug Effects," *New England Journal of Medicine,* November 23, 2006, pp. 2169–71.

Index